Between Mechanics and Architecture

Entre Mécanique et Architecture
edited by Patricia Radelet-de Grave and Edoardo Benvenuto

Towards a History of Construction
edited by Antonio Becchi, Massimo Corradi, Federico Foce and Orietta Pedemonte

Essays on the History of Mechanics
edited by Antonio Becchi, Massimo Corradi, Federico Foce and Orietta Pedemonte

Essays on the History of Mechanics

In Memory of Clifford Ambrose Truesdell
and Edoardo Benvenuto

Edited by Antonio Becchi, Massimo Corradi,
Federico Foce, Orietta Pedemonte

With the support of the Associazione Edoardo Benvenuto and
of the Fondazione Cassa di Risparmio di Genova e Imperia

Springer Basel AG

Editors:

Antonio Becchi, Massimo Corradi
Federico Foce, Orietta Pedemonte
Facoltà di Architettura
Stradone di Sant' Agostino, 37
16123 Genova
ITALY

Editorial Consultant:

Kim Williams
Kim Williams Books
Via Mazzini, 7
50054 Fucecchio (Florence)
ITALY
k.williams@leonet.it

Between Architecture and Mathematics: The Work of Clifford Ambrose Truesdell and Edoardo Benvenuto
International Symposium at Genoa, 30 November – 1 December 2001

Sponsored by: Accademia Ligure di Scienze e Lettere
                           Fondazione Cassa di Risparmio di Genova e Imperia
                           Università degli Studi di Genova
                           Facoltà di Architettura
                           Facoltà di Ingegneria
                           Dipartimento di Scienze per l'Architettura
                           Dipartimento di Ingegneria Strutturale e Geotecnica
                           Rotary Club della Città di Genova

With the patronage of: Comune di Genova

With the collaboration of: CJT Italia – Agenzia di Genova

Local Organizing Committee: Danila Aita, Giovanna Aita, Antonio Becchi,
                           Giovanni Benvenuto, Massimo Corradi,
                           Federico Foce, Orietta Pedemonte

A CIP catalogue record for this book is available from the Library of Congress, Washington D.C., USA

Bibliographic information published by Die Deutsche Bibliothek
Die Deutsche Bibliothek lists this publication in the Deutsche Nationalbibliografie; detailed bibliographic data is available in the Internet at <http://dnb.ddb.de>

ISBN 978-3-0348-9435-7   ISBN 978-3-0348-8091-6 (eBook)
DOI 10.1007/978-3-0348-8091-6

©2003 Springer Basel AG
Originally published by Birkhäuser Verlag in 2003
Softcover reprint of the hardcover 1st edition 2003
Printed on acid-free paper produced from chlorine-free pulp
Cover Illustration: Edoardo Benvenuto

ISBN 978-3-0348-9435-7

9 8 7 6 5 4 3 2 1   www.birkhäuser-science.com

Table of Contents

PREFACE

The history of mechanics, and more particularly, the history of mechanics applied to constructions, constitutes a field of research that is relatively recent. Be that as it may, it has attracted illustrious scholars, beginning in the nineteenth century with Saint-Venant, Todhunter and Pearson and continuing in the twentieth century with Duhem, Mach, Dugas, Timoshenko, Truesdell, Heyman, Szabó, Charlton, Benvenuto and Di Pasquale, to name only the greatest. In the last twenty years, the works of these authors have led new generations of scholars to delve into themes of research that involve diverse disciplines – from mathematics to construction, from architecture to geometry, from the strength of materials to the mechanics of solids and structures – leading to the formulation of the research project *Between Mechanics and Architecture*, begun by Edoardo Benvenuto and Patricia Radelet-de Grave on the occasion of the *XIXth International Congress of History of Science* (Saragozza 1993) and continuing with symposiums in Genoa (1996), Liège and Louvain-la-Neuve (1997) and Pescara (1998).

After the death of Edoardo Benvenuto (1940–1998) and Clifford Truesdell (1919–2000), the *Associazione Edoardo Benvenuto per la ricerca sulla scienza e l'arte del costruire nel loro sviluppo storico* (founded in 1999) decided to dedicate an international symposium to their contributions to the field of the history of mechanics. The close ties between the two scholars, born on the occasion of the publication of the book by Edoardo Benvenuto, *An Introduction to the History of Structural Mechanics* (New York: Springer, 1991), induced the *Associazione* to organize the meeting during which the papers in this volume were presented. The meeting took place in Genoa, 30 November – 1 December 2001, at the Faculty of Architecture, where Benvenuto taught "Structural Mechanics" until his death and where he was the for eighteen years. Participating in the symposium were Louis L. Bucciarelli (Cambridge, Massachusetts), Sandro Caparrini (Turin), Jean Dhombres (Paris), Jacques Heyman (Cambridge, UK), Santiago Huerta (Madrid), Karl-Eugen Kurrer (Berlin), Giulio Maltese (Rome), Gleb K. Mikhailov (Moscow), Patricia Radelet-de Grave (Louvain-la-Neuve), David Speiser (Basel) and Piero Villaggio (Pisa).

The present volume, which is part of the series *Between Mechanics and Architecture*, together with the recent publication *Towards a History of Construction* (Basel: Birkhäuser, 2002), is intended as an homage to the two eminent scholars who made a determinant contribution to the history of mechanics.

We thank the Faculties of Architecture and Engineering, the Departments of Sciences for Architecture and Structural Engineering and Geotechnics of the

University of Genoa, the Ligurian Academy of Science and Letters, the Foundation of the Cassa di Risparmio di Genova and the Rotary Club of the city of Genoa for having collaborated on the organisation of the symposium. We wish also to thank Kim Williams for her amenability and her patience in the preparation of the texts for this volume.

The editors
Genoa, 15 January 2003

TRUESDELL AND THE HISTORY OF THE THEORY OF STRUCTURES

Jacques Heyman[1]

There is an established hierarchy in the field of physical science: the mathematician tops the physicist who in turn tops the engineer. Further, the historian of science knows that he is operating on a higher plane than those whose history he is studying. This is not a view shared by working scientists, and Truesdell was aware of the contempt he was in danger of arousing by defecting from his proper work to study its history. But mathematics developed so rapidly that only a practising mathematician such as Truesdell, not a professional historian, can give a proper description of, for example, the work of Euler.

All engineers suffer from paranoia; but even though an engineer is paranoiac, he may in fact be low man on the totem pole. Certainly a physicist knows that he is measurably superior to the engineer – so much so that engineering problems are hardly worth the physicist's attention. Indeed, the problems are so trivial that they are, for the large part, invisible to the physicist. In the same way the mathematician knows that the problems of physics, if only they were properly formulated, could be solved without ugly recourse to real experiment, or without the need for virtual experiment by computer.

There is thus an established hierarchy in the field of physical science (although some mathematicians would prefer their subject to be thought of as a humanity) – the mathematician tops the physicist who in turn tops the engineer. One is reminded of the game where two children state simultaneously that they choose to be A, B or C. If A > B > C then there is a clear winner at each play of the game, or indeed a draw if each child happens to make the same choice – obviously the best strategy would be for both children to choose to be A. However, the game as actually played uses the curious (mathematically curious, that is) rule that, although A, B and C are different, A > B > C > A. *Scissors* will cut *paper; paper* will wrap *stone; stone* will blunt *scissors.* The paranoiac engineer struggles always, although he knows that he occupies the lowest position and so carries all above – indeed those above on the totem pole could not exist without the engineer's support – the engineer struggles to prove, if only to himself, that he is in some way superior to the mathematician, and certainly to the physicist. And indeed on occasion the engineer turns physicist, or mathematician, in order to gain information that would not otherwise be available.

---

[1] 3 Banhams Close, Cambridge CB4 1HX, UNITED KINGDOM

In parallel with this fictive pecking order in the physical sciences, there is another in the humanities – and all those working in the fields of the humanities may well consider themselves superior to all scientists. To be sure, the determination of an ordering of some sort of a philosopher, a musicologist and a theologian may be difficult, but a historian, say specifically a historian of science, will know that he is operating on a plane much higher than those whose history he is studying.

This is not a view shared by working scientists, and Truesdell, as applied mathematician, was well aware of the contempt he was in danger of arousing by defecting from his proper work to study its history [Truesdell 1984: vii]. When inspiration for research runs dry, then the scientist may 'dwindle into a philosopher of science'. At the same time the scientist, like Truesdell, may well confess to being 'untrained in classroom philosophy'; thus the renegade scientist will be viewed with suspicion both by the working scientist and by the trained historian. However, it is a fact that (to take but one example) the professional historian is able to give a coherent account of the work of Newton (and a fine model is given by the work of Westfall [1980]), but mathematics developed so rapidly that the professional historian cannot give a proper description of work in the mid-eighteenth century – of the work of Euler, for example. There is thus an immense lack of 'professional' historical study of this period – a lack so spectacularly made good by the work of Truesdell himself.

Truesdell came late to the history of the mechanics of solids; when he reviewed Timoshenko's *History* [1953] he had not yet read carefully Saint-Venant's fundamental study of 1856 [Truesdell 1984: 252]. Yet by 1960 he had what seems to be a complete grasp of the history of *The rational mechanics of flexible or elastic bodies*, as demonstrated by the volume of that title published as part of Euler's *Opera Omnia*. (Timoshenko's full title is *History of strength of materials, with a brief account of the history of theory of elasticity and theory of structures*; these three (separate) topics are distinguished below.) Timoshenko had read Saint-Venant [1864], and also Todhunter and Pearson [1886, 1893], and had added material from the next 75 years or so; there are still large gaps, but Truesdell found himself unable to put the book down. Truesdell in his own book published seven years later [1960] makes good the gaps; at the same time the emphasis is altered subtly. As acknowledged by Truesdell, Timoshenko was addressing engineers, and he was

> relying heavily on the scantness of American engineers' education
> ... although [Timoshenko's] books [including his other technical
> books] are almost totally devoid of originality, they served to
> acquaint American mechanical and civil engineers with theory and

history they were otherwise unlikely to encounter [Truesdell 1984: 253].

Truesdell, on the other hand, is specifically concerned with mathematical theory, and with the theory of elasticity.

Truesdell's dismissal of American engineers as unlikely to know much about the theory and history of engineering is, of course, justified, and the remark could have been made about engineers of any nationality. Indeed, Benvenuto makes the same point: 'If we asked an engineer about the origins of the equations he or she uses constantly, the reply would be disappointing. They exist; nothing else matters. Why be curious about their derivation?' Engineers are not interested in the background to their subject, and it cannot be emphasized too strongly that, although mathematicians, physicists and engineers appear, to the layman, to use the same language, they are, each of the three, pursuing different objectives. The layman's confusion arises from the fact that he does not understand the language being used – that is, the mathematics, and to some extent the science, employs an encryption to which the layman has no key. The mathematician masters the code, and manipulates it, to create a logically interconnected system of thought – a system, moreover, which is capable of throwing up totally new ideas. These ideas need have no relevance at all to anything in the real world, but this does not detract from their inner power nor (for the mathematician) their beauty. And, from time to time, a result is obtained which is powerful in the real world – the study of prime numbers leads to the creation of virtually unbreakable military codes; Euler's study of the differential equation of the elastica leads to a formula for the buckling of columns.

The physicist, on the other hand, always stays closer to reality. Whether experimental or theoretical, he wishes to establish the laws of 'Nature' which govern his world. His approach is 'scientific' and not metaphysical – as a physicist, he does not wish to discuss the 'yellowness' of the colour yellow, but to assign a numerical wavelength to the colour yellow in the spectrum. The physicist uses the coded language of mathematics in order to deepen understanding of a particular branch of his science.

By contrast, the engineer uses the same language in order to do something – whether it be to create an electronic circuit, a jet engine or a skyscraper. For this he selects appropriate mathematical or physical tools and, as has been noted, as an engineer he is not interested in the theory which went into the making of those tools, nor yet in the history of that theory, but merely in whether or not they will do a particular job.

The boundaries between mathematics/physics/engineering are, of course, not rigid. The engineer may find that no tool exists to perform a particular task – but nevertheless the task must be done. The engineer is not playing a game; he cannot walk away and pursue some other more tractable problem. A solution must be found, and the engineer must turn physicist and make experiments, or mathematician and evolve new theories, if a suitable tool is to be devised. These three types of activity have led to three main topics in the study of what is grandiosely called *The mechanics of solids*; the structural engineer's main activities are concerned with *The theory of structures*; the physicist studies *The strength of materials*; and the mathematician's chief interest is in *Elasticity*. The titles of these three topics are sometimes used interchangeably to describe the technical activity which lies behind, but the topics are in fact quite distinct (with, as usual, some blurring at overlapping boundaries).

These divisions did not exist before, say, Galileo in the seventeenth century (although antecedents may be found). Structural design codes did indeed exist, and had existed for the previous 2000 years, and these codes, from the book of Ezekiel through Vitruvius to the secret books of the Masonic lodges, gave efficient and correct rules for the design of great masonry structures – Greek temples, Roman aqueducts, Gothic cathedrals. The rules were, essentially, rules of proportion – that the height of a column should be a certain multiple of its diameter, for example. A column designed in this way had been found by experience to be satisfactory – it could play its part in a complete building without buckling or crushing; moreover, experience had shown that the column could be built to any scale, contrary to 'modern' intuition.

Ratios – that is, proper fractions – have some mathematical interest, but their study will not detain the professional mathematician for long. However, design by ratios, at any scale, is of great interest to the structural engineer, precisely because it now seems counter-intuitive. The rules worked because the materials used for ancient building – stone and to some extent wood – are very lightly stressed in the completed structures. Very large-span stone bridges could be constructed, and cathedrals built to a staggering height; all that was necessary was that the shape (of an arch ring, of a flying buttress) should be correct, and this is a geometrical requirement, to be satisfied by following the rules of proportion. (There was, however, one mathematical problem, known to the ancient Greeks, the Romans and to medieval masters – some dimensions, of importance on a building site, cannot be expressed as proper fractions, and so could not be measured on a marked scale of length. The Pythagorean demonstration that the square root of two is irrational is a mathematical milestone.)

Galileo seems to be the first to have attacked the ancient and medieval rules for structural design. He exposes the inexorable operation of the square/cube law

– it is in fact impossible to build temples to an enormous height, and an elephant is at about the limit of size of land-based animals. The basic problem considered by Galileo in the first of his 1638 *Two new sciences* is the fracture of beams: if the breaking strength in tension is known for a certain material, how much load can be carried by a cantilever beam made from the same material? There is no question here of notions of *elasticity*, and although Galileo elsewhere just glimpses the difficulties associated with hyperstatic structures (a beam on three supports), the cantilever problem is not one lying in the field of *theory of structures*. It is an applied problem in *strength of materials*. Needless to say, Galileo tackles the problem brilliantly, and his solution is the foundation for the development of design methods for structures working their new materials (iron, and then steel and reinforced concrete) much closer to their limits of strength.

'Galileo's problem' received much attention throughout the eighteenth century, with attempts being made to modify the theory to accord more closely with the results of practical tests. The problem was still treated as one of strength of materials; only very occasionally was it appreciated that the results could not be applied directly to (for example) a beam on several supports. There was very little mathematical interest in these analyses; indeed, the giants (e.g., Leibniz, or James Bernoulli), although they took up the problem, found nothing to engage their attention. However, a much greater problem was posed by the elastica (and here the giants were indeed engaged) – to find the shape taken by an elastic strip when subjected to specified end displacements. This was the tremendous challenge met by Euler in 1744; his handling of the hideously non-linear fourth-order differential equation is almost unbelievable (see, for example, [Truesdell 1960: 199ff]). Moreover, Euler himself saw that for practical applications (to a beam, say, in a building, for which deflexions would be small, or to a column in compression) the basic equation could be reduced to a second-order linear equation, capable of easy solution. Galileo had not considered the *shape* adopted by his loaded cantilever (although others had made guesses); now, a hundred years later, elastic deformations could be calculated for simple structural elements.

Thus simple elasticity had been discovered. It was to be a further hundred years, however, with the formulation of the full three-dimensional stress tensor, before the subject proper of elasticity was born, say with Saint-Venant in the 1850s. The subject offered immediate challenges for the applied mathematician – even simple problems in two dimensions might lead to eighth-order partial differential equations, while only a tiny handful of three-dimensional solutions could be obtained in closed form. For over 150 years the subject has occupied fine mathematical minds, and, with the complexities of anisotropy and non-linearity, it will continue to do so.

Early in the nineteenth century, however, elastic ideas found their way into the analysis of building structures. The century and a half since Galileo had confirmed that his solution for the strength in bending of a beam was essentially correct – a beam twice as deep was four times as strong, for example – but the numerical constant giving the actual bending strength seemed to vary according as the material was stone or wood, if experimental results were to accord with theory. However, from this work emerged gradually the realization that *strength of materials* was something different from *theory of structures*. The value of Galileo's bending moment which caused fracture of the beam could be found by theory using the equations of statics; those same equations could not, however, determine the bending moments in, for example, a beam on three supports (or in Galileo's beam if it had an additional prop).

Navier, in his published *Leçons* [1826], saw his way through this difficulty, and must be regarded as the first founder of the theory of structures. Since the equations of statics do not furnish enough information to solve a real (hyperstatic) structural problem, extra information must be found elsewhere. One calculation that could be made – from the work of Bernoulli/Euler – would determine the shape of a bent beam, and so a whole new class of statements was available dealing with (small) elastic deformations, to supplement the equilibrium statements. Finally, the deformed beam must obey certain boundary conditions – a beam on three supports must continue to remain in contact with those supports – and a set of compatibility statements of this sort was needed. Navier also applied these three master ideas of static *equilibrium*, of (elastic) *material properties* and of *compatibility* of deformation to the trussed framework, and he showed that the three master statements always generated enough equations to provide a solution to any structural problem. It was to be a century before it was realised that the Navier solution did not correspond to the real behaviour of a practical structure.

Navier's approach is straightforward, but involves an enormous number of equations. As an example, three equations of static equilibrium will suffice to determine the forces in the legs of a three-legged table. Only the same three equations, however, may be written for the usual (hyperstatic) four-legged table, and the Navier solution requires a knowledge of the flexibility of the table top, of the compressibility of the legs, and so on. Only when all this information is assembled can the leg forces be determined. The difficulties of calculation were quickly appreciated, and the second half of the nineteenth century saw the development of ingenious methods of evaluating the equations (essentially, of solving large numbers of simultaneous equations), by successive approximations, for example, by graphical methods, and by making experiments on models (not necessarily slavish scale models). Some of this work has an authentic intellectual kick, but, in the final analysis, none has a deep mathematical content. An exception lies in the development of energetic principles, exemplified by the theorems of Castigliano. By and large, the field of *theory of structures* has been left, for the last two centuries, to the engineer. Mathematicians and physicists working in the field of mechanics of solids have been concerned with the other two disciplines. Navier had, apparently, solved the structural problem; working out results for practical design was mere mechanical drudgery.

The drudgery has, of course, been taken over by the electronic computer. It is no longer necessary to put up with approximate solutions for structural problems – the nineteenth-century equations can now be solved, if not in closed form, at least to a fantastic accuracy, far beyond the needs of a practical designer. Those needs may be recapitulated in summary. The essential problem of the *theory of structures* is to determine the way a hyperstatic structure carries its load (to find the forces in the four legs of a table). Once these internal stress resultants have been calculated, then the science of *strength of materials* may be used to proportion the members so that they have adequate strength. In practice, this process may be iterative, and it involves the use of the three master statements of *equilibrium, material properties* and *compatibility.* The engineer is not necessarily interested in the actual deformations of a structure, which in practice may be very small – the prime structural problem is to determine the internal stress resultants (the forces in the legs of the table). The Navier algorithm (or algorism, as Truesdell would say) gives a solution to this problem, and seems so evidently correct that it is perhaps not surprising that it was about a hundred years before anyone thought to try to measure the actual behaviour of a real structure.

The first full-scale tests on building frames were made in the 1920s in London, and the results were, in the fullest sense of the word, astounding. The internal stress resultants (in this case, bending stresses in beams) that were measured in practice bore almost no relation to those confidently calculated by the elastic (Navier) designer. The investigators quickly found what was wrong; while the equilibrium equations were obeyed, and the elastic equations of deformation were found to be almost exact, the boundary conditions used in the analysis did not represent reality. A small error in manufacture, a tiny dimensional mismatch, an inevitable forcing together of members during construction – all these led to a great change in the values of internal stresses.

To revert to the simple example of the four-legged table, it is true that the use of elastic finite elements will, with the aid of a computer, lead to precise predictions of the forces in the four legs. However, a real, nearly rigid table, placed on a real, nearly rigid floor, will rock; if one leg of the table is off the ground by a fraction of a millimetre, then the force in that leg is zero, despite the prediction of the computer program. The programme has assumed, unthinkingly, that the floor is uniform, and that all legs start and remain in contact – these are the boundary conditions. If an accurate description of the floor's irregularities had in fact been given to the computer, then the program could have confirmed that a particular leg carried no load. However, that is still not the solution to the real structural problem – the table is placed randomly on a slightly irregular floor, so that any one of the four legs may be marginally clear of the ground. This is the actual problem facing the engineer – how is the structure to be designed if the boundary conditions are unknown (and indeed unknowable), so that the third of Navier's statements cannot be used?

Euler appreciated in 1773 the difficulties of analysis of hyperstatic structures and, specifically, he contrasted the behaviour of a three-legged table with that of the hyperstatic table with four legs. He was aware explicitly that one of the four legs might, if the table were on an uneven floor, be carrying zero load, but he was unable to make progress with the problem. Others gave it their attention during the next half century, with equal lack of success. Navier's solution in 1826 was a revelation – in Benvenuto's words, the goal had been obtained. It seems extraordinary that no-one remarked that the actual problem had not been addressed – the determination of the forces in the four legs of a table placed on an uneven floor. Navier assumed precisely that the floor was even.

The *plastic theory* of structures deals with the real structural problem, and with hindsight it might be thought that it was devised for that purpose. In fact, plastic theory was developed in the first instance to deal with the particular problem of the design of steel frames – of those frames whose real behaviour departed so radically from the predictions of elastic designers. The development followed obvious traditional lines – only three kinds of statement can be made in the theory of structures, and those statements were modified so as to be seen from a 'plastic' rather than an 'elastic' viewpoint. The basic equations of static *equilibrium* are paramount, and hold unchanged. The use of material properties is greatly simplified – instead of a linear elastic stress/strain law, no mention is made of strain in simple plastic theory. Instead, the material is supposed to be capable of carrying any stress (or, sophisticatedly, combination of stresses) below an experimentally determined *yield limit*; at that limit, the material is supposed to be sufficiently ductile to sustain any reasonable displacement. Finally, no precise statement is made about boundary conditions – the floor supporting the legs of the four-legged table can have any small unknown irregularities.

That it was possible to mysteriously reject this last of the three master structural statements was discovered nearly 100 years ago. Kazinczy tested some fixed-ended beams in order to find experimentally what degree of fixity was needed for each beam to develop its full strength (it was known that very small changes in fixity would alter markedly the elastic stresses). Kazinczy found that the abutments had merely to be sufficiently strong to sustain the maximum moments developed in the beams, and could be allowed to deform to any small degree. Similarly, experiments some 20 years later confirmed that collapse loads of continuous beams on several supports were quite independent of small random settlements of those supports.

Thus the examination of the plastic behaviour of structures involves the construction of solutions using only the equations of static *equilibrium* together with a knowledge of the *yield limit* of the material. This major development in the theory of structures is unremarked by Timoshenko. He does indeed just mention plasticity, and notes that Saint-Venant had tackled some problems. These problems, however, all lie in the field of strength of materials – for example, the estimation of the value of full plastic moment in Galileo's beam, or discussion of the plastic torsion of non-circular shafts. Truesdell seems not to refer in any of his writings to any plastic problem. To be sure, Timoshenko in 1953 could not have been aware of the implications for structural theory of the ideas of plasticity; Truesdell, however, may also have had a more fundamental reason for his neglect.

It is perhaps unfortunate that 'plastic design', with its basic implication that a collapse analysis is being made, has come to be accepted as a description of the method. The method in fact deals with the equilibrium of a structure in its working state, and the rock on which it is founded is the so-called 'safe theorem'. The theorem can be stated simply. There is no single state of a hyperstatic structure which can be observed in practice (it is not known which one of the four legs of the table is off the ground) – nevertheless, if the structural engineer can find any one set of internal forces for the structure with which it is comfortable (equilibrium and the yield condition are satisfied) then this is proof that the structure is safe. That is, there is no possible alternative equilibrium state which would imply collapse of the structure.

This plastic theorem for a structure makes no reference at all to elastic properties of the material. The splendid mathematical developments in elasticity which occurred after 1826 have at best only marginal relevance to the activity of the structural engineer; working deflexions may perhaps be calculated roughly, but elastic theory is of no help in estimating the final strength of a ductile structure. Truesdell was immensely erudite across many fields, not only elasticity, but he appears to have had no understanding of the intensely interesting problem of hyperstatic structures. His attitude is perhaps exemplified by his denigration of Pearson. Pearson, a professor of applied mathematics, put together, after Todhunter's death, the massive 2000-page *A history of the theory of elasticity* [Todhunter and Pearson 1886, 1893]. Truesdell is particularly upset that Pearson's frontispiece to this work shows drawings of rupture-surfaces of cast iron – and this in a book whose title uses the word *theory*. Truesdell's interest lay in the mathematics of the theory of elasticity; by contrast, Pearson was aware that the objective of the whole science was to analyse, and design, real constructions. Truesdell seems to have had a complete understanding of all aspects of elasticity theory, but it was for him a closed science, without application.

It was, however, someone working as a mathematician, Gvozdev, who established in 1936 the fundamental theorems of the plastic theory of structures (including the safe theorem quoted above). The work was not published until 1938, and then obscurely, and in Russian. In the following decade engineers had mastered, mechanically, the essentials of plastic design, and had even, in 1948, altered the building codes in the United Kingdom to permit officially the new method. But they had done this without any firm theoretical framework; they had been unable to assume the role of mathematician to create for themselves the powerful theorems of Gvozdev, of which they were ignorant. Perhaps they would have done so in time – as it is, the new structural theory was established from about 1950 as the result of contributions from mathematicians, physicists and engineers. Perhaps this is an example of the true resolution of the mathematically impossible A > B > C > A.

Bibliography

BENVENUTO, E. 1991. *An introduction to the history of structural mechanics. Part I: Statics and resistance of solids. Part II: Vaulted structures and elastic systems.* 2 vols. New York: Springer.

GVOZDEV, A.A. 1938. The determination of the value of the collapse load for statically indeterminate systems undergoing plastic deformation. P. 19 in *Proceedings of the conference on plastic deformations*, December 1936. Akademiia Nauk, Moscow/Leningrad. In Russian; Engl. Trans. by R. M. Haythornthwaite, *International Journal of Mechanical Sciences*, vol. 1 (1960): 322-335.

HEYMAN, J. 1998. *Structural analysis : a historical approach.* Cambridge: Cambridge University Press.

KAZINCZY, GÁBOR. 1914. Test with clamped beams (in Hungarian). *Betonszemle*, vol. 2: 68-71, 83-87, 101-104. (For an English summary of the paper, see S. Kaliszky, Gábor Kazinczy 1889-1964, *Periodica Polytechnica*, vol. 28: 75-93, Budapest, 1984.)

NAVIER, C. L. M. H. 1826. *Resumé des leçons données à l'École des Ponts et Chaussées, sur l'application de la mécanique à l'etablissement des constructions et des machines.* Paris. (2nd edition, 1833; 3rd edition, with notes and appendices by B. de Saint-Venant, 1864.)

SAINT-VENANT, B. DE. 1864. Historique abrégé de recherches sur la résistance et sur l'élasticité des corps solides. Pp xc-cccxj in *Resumé des leçons données à l'École des Ponts et Chaussées, sur l'application de la mécanique à l'etablissement des constructions et des machines*, 3rd ed. Paris.

TIMOSHENKO, S. P. 1953. *History of strength of materials.* New York and London: McGraw-Hill.

TODHUNTER, J. AND K. PEARSON. 1886-1893. *A history of the theory of elasticity and of the strength of materials.* 2 volumes in 3 parts. Cambridge: Cambridge University Press. (Repr. New York: Dover, 1960.)

TRUESDELL, C.A. 1960. *The rational mechanics of flexible or elastic bodies 1638-1788.* Introduction to *Leonhardi Euleri Opera Omnia*, second series, vol. XI (2). Zürich:Orell Füssli.

———. 1964. *Essays in the history of mechanics.* New York: Springer.

———. 1984. *An idiot's fugitive essays on science.* New York: Springer.

———. 1987. *Great scientists of old as heretics in 'the scientific method'.* Charlottesville: University of Virginia Press.

WESTFALL, R.S. 1980. *Never at rest.* Cambridge: Cambridge University Press.

Clifford Ambrose Truesdell

DEVELOPMENT OF STUDIES IN THE HISTORY
OF ELASTICITY THEORY AND STRUCTURAL MECHANICS

Gleb K. Mikhailov[1]

The development of the History of Rational Mechanics covers about 150 years only. However, historical studies at the end of the nineteenth and the beginning of the twentieth century were usually restricted to the principles of the Mechanics of Discrete Systems. The first professional review of the history of elasticity theory was given by Adhémar Barré de Saint-Venant in his *Historique abrégé des recherches sur la résistance et sur l'élasticité des corps solides* (1864). Another exception is the *History of the theory of Elasticity and of the Strength of Materials from Galilei to Lord Kelvin* (1886-1893) by Isaac Todhunter and Karl Pearson, which presents on 2200 pages summaries of most of the works on selected topics. During the nineteenth century, Mechanics was mainly considered to be a part of Applied Mathematics. A significant change occurred at the beginning of the twentieth century when Continuum Mechanics acquired special significance. In the middle of that century, Stephen Timoshenko published the *History of Strength of Materials with a brief Account of the History of Theory of Elasticity and Theory of Structures* (1953). Although this historical essay was extremely informative, Clifford Truesdell was the first to begin in the 1950s broad and profound investigations into the History of Continuum Mechanics. First of all, his work is distinguished by a thorough knowledge of the sources. He was perhaps the only famous, creatively working scientist in Rational Mechanics who was professionally engaged in the History of Mechanics as well, and he can be rightly considered the founder of its modern trends. The history of structural mechanics was treated in the spirit of Truesdellian fundamentals later by Edoardo Benvenuto in his numerous studies and, particularly, in *An Introduction to the History of Structural Mechanics* (1991).

The beginnings of the history of mechanics

The history of mechanics emerged comparatively recently as an independent discipline. It covers altogether about 150 years, prior to which there was hardly any systematic accumulation of historical material in this field of study. In fact, discussions of the principles of mechanics, frequently associated in the eighteenth century with priority arguments, can be found in some older publications. Thus, surveys of the development of the principles of statics and dynamics are already contained in Louis Lagrange's (1736–1813) *Mécanique analytique* (1788). Actually, right up to recent times, many subsequent authors found their material in this treatise, with all its merits and defects. "Lagrange's histories usually give the right references but misrepresent or slight the contents," was a comment on

[1] Secretary-General, Russian National Committee on Theoretical and Applied Mechanics, Vernadskogo Ave. 101, Moscow 117256, RUSSIA

this state of affairs by Clifford Truesdell, a severe critic of Lagrange's *Mécanique analytique* [Truesdell 1968: 247].

There are also some remarks on mechanics in many old books dealing with the history of mathematics, for example, in the well-known *Histoire des mathématiques* by Jean Etienne Montucla (1725–1799) completed later by Joseph Jérôme Lalande (1732–1807) [Montucla 1802], but such surveys mainly enumerate sources without a proper analysis.

In the middle of the nineteenth century, mechanics attracted the attention of broad circles of natural philosophers. The discovery of the Law of Conservation of Energy and attempts to elaborate a unified mechanistic description of the world required an understanding of the basic concepts of mechanics. In this connection, the first step was an attempt to comprehend (or reassess) the laws of mechanics of Newton's *Principia*. The fundamental difficulties of a formal axiomatization of the foundations of mechanics (and, in particular, definitions of the concepts of mass and force) led to a broad and, at times, sharp discussion in which, to one degree or another, practically all of the most important scientists of the time partook.

In connection with the interest displayed in this range of problems, the Philosophical Department of Göttingen University proposed in 1869 a contest for the best analytical essay devoted to a critical history of the basic principles of mechanics. It gave a clear-cut definition of the requirements, relating to the historical part of the investigation (by the statement: "when and by whom and in what connection with specific problems was each separate essential principle of Mechanics first found and enunciated") as well as to an analysis of a logical and experimental substantiation of the principles and their interrelationship with philosophical theories. Altogether five works were submitted. The first prize went to Eugen Dühring (1833–1921) for his *Kritische Geschichte der allgemeinen Principien der Mechanik* [1873]. The next work of this kind was Ernst Mach's (1838–1916) *Die Mechanik in ihrer Entwickelung historisch-kritisch dargestellt* [1883]. The philosophical sections of these and other works of Dühring and Mach initiated an extensive discussion, which is still evident today. In fact, the historical aspects of these studies (more extensive in the case of Dühring) did not attract any particular attention.

The historical part of Mach's *Mechanik* was subjected to criticism by Truesdell [1968: 85–87], whose interest was aroused by the literature's frequent use of quotations from Mach's book. His criticism is that Mach was acquainted only superficially with the source material.

A number of problems in the history of mechanics were treated at the end of the nineteenth and the beginning of the twentieth centuries in works devoted to the general history of physics and mathematics. However, in such studies, the discussion of mechanics was usually restricted to the principles of Newton's mechanics and was mostly not of a deep nature. An illustration of this are the four volumes of *Die Geschichte der Physik in Grundzügen* [1881–1890] by Ferdinand Rosenberger (1845–1899), in general an informative book.[2]

A new trend in the history of mechanics was initiated at the start of the twentieth century by Pierre Duhem (1861–1916), who acquainted modern science with the mechanics of the Middle Ages and about whom Truesdell wrote, "... the great man who founded the modern history of science ... A scientist, as a sound historian of science must be, he has left permanent achievements in physical chemistry, thermodynamics, hydrodynamics, and elasticity" [Truesdell 1968: 25]. Subsequently, Truesdell has emphasised more than once the necessity that only a real scientist in the topic must write its history.

Duhem wrote such fundamental monographs as *Les origines de la statique* (2 volumes, 1905–1906), *Etudes sur Léonard de Vinci* (3 volumes, 1906–1913) and *Le système du Monde* (10 volumes,1913–1959), in which he presented the first analysis of medieval views regarding statics and dynamics. Later on, a study of the medieval sources appeared in the detailed works of Alexander V. Koyré (1892–1964), Eduard J. Dijksterhuis (1892–1965), Ernest A. Moody (1903–1975), Annelise Maier (1905–1971) and Marshall Clagett.

During the first half of the twentieth century, it is difficult to detect any new schools or trends in the history of mechanics, with the exception of the investigations into the mechanics of the Middle Ages already referred to.

During this whole period, there was – in many countries – a great deal of activity in analysing the work of individual scientists working in mechanics, in particular in relation to anniversaries and publication of various *Collected Works*. The main concern was collection of material, which simplified the work of subsequent historians in critical analyses of separate branches of mechanics.

A more profound interest in the history of mechanics has occurred everywhere during the last fifty years. At the start of this period, there appeared four fundamental publications. Two monographs devoted to medieval statics and dynamics: the collection *The medieval science of weights* (*Scientia de ponderibus*) which includes treatises ascribed to Euclid, Archimedes, Thabit ibn Qurra, Jordanus de Nemore and Blasius of Parma edited by Ernest A. Moody and Marshall Clagett [1952], and the subsequent treatise *The science of*

---

[2] However, Rosenberger's book mentions neither Louis Navier (1785-1836) nor the foundations of the hydrodynamics of a viscous fluid.

Mechanics in the Middle Ages by Marshall Clagett [1959]. The other two publications were large monographs by René Dugas (1897–1957) – *Histoire de la Mécanique* [1950] and *La Mécanique au 17e siècle* [1954]. These two voluminous books quote huge material and can be used as a starting point for further original investigations. However, as Truesdell has stressed, Dugas's analysis is not always correct.

As regards the general investigations in the history of mechanics up to the end of the twentieth century, the majority has dealt with the principles of Newtonian mechanics, the Variational Principles and their interrelationships.

However, it is a curious fact that, despite the tremendous range of literature devoted to Isaac Newton over the past two and a half centuries, the principal fundamental studies of his scientific heritage have appeared during the last thirty to forty years. They include the first critical edition of Newton's *Principia* by Alexander V. Koyré and I. Bernard Cohen (1971–1972), a detailed analysis of the prehistory of the *Principia* in the work of Newton himself by John W. Herivel [1966], a special volume (1974) devoted to the *Principia*, in the 8-volumes edition *The Mathematical Papers of Isaac Newton* (1967–1981) by Derek T. Whiteside, *The preliminary manuscripts for Isaac Newton's 1687 'Principia': 1684–1685* (1989), and the seven volumes of *The Correspondence of Isaac Newton* (1959–1977).

First studies in the history of elasticity theory (Barré de Saint-Venant, Isaac Todhunter and Karl Pearson)

The development of continuum mechanics as a part of rational mechanics did not attract so much attention in old historical essays. As a rule, it was considered only in aspects of some parts of applied mechanics such as hydraulics and strength of materials. Nevertheless, one can find some extensive remarks on the history of theoretical hydraulics in the *Histoire des mathématiques* [Montucla 1802].

Perhaps, the first historical review in the field of the theory of elasticity was in the *Traité analytique de la résistance des solides* [1798] by Pierre Simon Girard (1765–1836). "It is not only valuable as containing the total knowledge of that day on the subject, but also by reason of an admirable historical introduction", subsequently verified by Todhunter and Pearson [1886: 74].

In the field of the history of structural mechanics and arch theory, one should also mention a short historic-bibliographical note [1843] by a German engineer Wilhelm Lahmeyer (1818–1859) and the historical survey [1852] by the well-known French engineer and scientist Jean Victor Poncelet (1788–1867), about which Todhunter and Pearson wrote, "This is a very valuable criticism of the various theories of the arch propounded up to 1852 ... The paper forms a most interesting historical resumé of the subject" [1893, pt. 1: 678-679].

The next, much larger professional historical review in the field of the theory of elasticity and strength of materials, is due to the famous French scholar Adhémar Barré de Saint-Venant (1797–1886) who was probably the greatest scientist in the field of continuum mechanics in the middle of the nineteenth century.[3]

His contribution to the theory of elasticity is very well known, as well as his equations of the unsteady flow in open channels. However, not all scientists are aware of the fact that he initiated the hydrodynamics of a viscous fluid. Unfortunately, his main Mémoire on this topic (of 1834) was lost by the official reviewer of the Paris Academy of Sciences and only a small note in the Paris *Comptes Rendus* of 1843 reminds us about it. As regards Saint-Venant's historical commentaries on the theory of elasticity and strength of materials, he gave them as a special appendix to the third edition of Navier's *Résumé des Leçons sur l'application de la mécanique à l'établissement des constructions et des machines* [1864] under the title *Historique abrégé des recherches sur la résistance et sur l'élasticité des corps solides*. Todhunter and Pearson wrote in this context,

> *The essential feature of scientific history is the recognition of growth, the interdependence of successive stages of discovery. This evolution is excellently summarised in Saint-Venant's* Historique [Todhunter and Pearson 1893, pt. 1: 106].

Later on, Saint-Venant also included some historical remarks in his voluminous commentaries to the French edition [1883] of Alfred Clebsch's (1833–1872) *Theorie der Elasticität der festen Körper* [1862] (Saint-Venant's commentaries – about 460 pages – is equal in length to the Clebsch's original text!).

The public life of Saint-Venant was not easy, but he worked intensively for almost 60 years. Announcing his decease, the President of the Paris Academy of Sciences Vice-Admiral Edmond Jurien de La Gravière (1812–1892) said,

> *La vieillesse de notre éminent Confrère a été une vieillesse bénie. Il est mort plein de jours, sans infirmités, occupé jusqu'à sa dernière heure des problèmes qui lui étaient chers, appuyé pour le grand*

---

3 Serious attention, particularly of historians of science, to the work of Saint-Venant has only been attracted recently, particularly by a special Symposium at Louvain-la-Neuve in July 1997 (its Proceedings have not yet been published). An interesting paper has been dedicated to Saint-Venant by Edoardo Benvenuto [1997].

passage sur les espérances qui avaient soutenu Pascal et Newton
(Comptes Rendus 1886, 102 : 73).[4]

A genuine encyclopaedia of the history of theory of elasticity and strength of materials appeared by the end of the nineteenth century in two volumes (published in three books): *History of the Theory of Elasticity and of the Strength of Materials from Galilei to Lord Kelvin* (1886–1893), edited and completed by Karl Pearson (1857–1936) on the basis of a manuscript of Isaac Todhunter (1820–1884) [Todhunter and Pearson 1886-1893]. In more than 2200 pages, it gives summaries of most of the work on selected topics for over two centuries and, although it is not exhaustive, it remains to this day an excellent reference work. No other division of mechanics has anything which is comparable with the fundamental compendium of Todhunter and Pearson.

However, they gave themselves only a modest evaluation of their work, "Our own 'history' is only a bibliographical repertoire of the mathematical processes and physical phenomena which form the science of elasticity, as a rule for the purpose of convenience chronologically grouped" [Todhunter and Pearson 1893, pt. 1: 106]. The work of Todhunter and Pearson did not initially attract much attention of the scientific community, but the understanding of its significance grew continuously, and it is now considered to be a heroic deed of its editors.

As to Isaac Todhunter himself, he was an extremely many-sided personage and scientist, as well as a diligent writer. In a brilliant review, Professor William Johnson assessed him as a "textbook writer, scholar, coach and historian of science" [Johnson 1996]. In fact, in addition to many textbooks, Todhunter compiled four significant historical fundamental research compendia in various branches of Mathematics and Mechanics: *Calculus of Variations* (1861, 532 pp.), *Mathematical Theory of Probability* [1865f, 624 pp.), *Mathematical Theories of Attraction and the Figure of the Earth* (1873, 2 vols., 984 pp.) and *Theory of Elasticity and the Strength of Materials* (1886–1896, 2 vols., 244 pp.). Each of these works contains extensive summaries of the entire, relevant world literature. As an other example of Todhunter's striking diligence, one can mention his 450-page account of the life and writings of the well-known philosopher and science historian William Whewell (1794-1866) [Todhunter 1876].

---

[4] Karl Pearson published a large excerpt from the *History of the Theory of Elasticity* devoted to Saint-Venant shortly after his death as a separate book under the title *The Elasticity Researches of Barré de Saint-Venant* (1889).

Further devolopment of studies in the history of continuum mechanics and the work of Stephen Timoshenko

An item of interest for the history of mechanics in the nineteenth century is the *Encyclopädie der mathematischen Wissenschaften,* the publication of which started in Germany in 1898 [Klein and Müller 1907-1914].[5] It comprises many independent extensive surveys with comprehensive bibliographical references which today render valuable auxiliary historical material. According to the original plan of the editors, the *Encyclopaedia* had to give a review of the modern state of science, including not only the so-called pure mathematics, but also applications in mechanics and physics, astronomy and geodesy, various branches of technology and other fields; moreover, the *Encyclopaedia* was "to prove with accurate references to the literature the historical development of the mathematical *methods* since the beginning of the nineteenth century".

The fourth volume of the *Encyclopaedia* (1901–1914) was devoted to mechanics[6] and edited by Felix Klein (1849–1925) and Conrad Müller (1878–1953), both of whom had a wide interest in the History of Mathematics.

The volume contains thirty-three surveys in various branches of Mechanics, several of which were written for this edition by outstanding scientists. The third part of the volume (Bd. IVC) includes nine surveys of the mechanics of deformable bodies (elasticity and strength of materials) and occupies 770 pages. They cover elasto-statics, elasto-dynamics, strength theory, theory of earth pressure, structural mechanics, and some general approaches to continuum mechanics. Various aspects of the theory of elasticity were reviewed by Theodor von Kármán (1881–1963), Horace Lamb (1849–1934), Conrad Müller (1878–1953), Aloys Timpe (1882–1959) and Orazio Tedone (1870–1920). The state of Structural Mechanics was discussed there by Martin Grüning (1869–1932) and Karl Wieghardt (1874–1924).[7] Unfortunately, the material of the

---

[5] This Encyclopaedia was also published, in collaboration with French scientists, since 1904 in France as *Encyclopédie des sciences mathématiques pures et appliquées* (Paris: Gauthier-Villars); however, due to World War I, the French edition ceased with only a few issues having been published.

[6] The publication of the volume was not finished properly due to World War I and the Index Issue of the volume appeared only twenty years later (1935).

[7] Among other authors of the volume were such eminent scientists as Paul Appell (1855-1930), Carl Cranz (1858-1945), George Darvin (1845-1912), Aleksey Krylov (1863-1945), Augustus E.H. Love (1863-1940), Richard von Mises (1883-1953), Hans Reissner (1874-1967). It is interesting that the surveys on hydrodynamics were written by A.E.H. Love, whereas the elasticity theory was reviewed by H. Lamb, although both of them are now mainly known as authors of famous treatises in the theory of elasticity and hydrodynamics, respectively.

Encyclopaedia has not used sufficiently in subsequent historical studies of the history of mechanics.

During the nineteenth century, mechanics was mainly considered to be a part of applied mathematics. A significant change in the position of mechanics occurred in the first quarter of the twentieth century, when various problems of continuum mechanics (including fluid and solid dynamics) became the topical problems of the time. It was reflected in the 1920s in the organisation of the new journal, *Zeitschrift für angewandte Mathematik und Mechanik* (1921), and the initiation of the International Congresses of Mechanics (starting from 1926). In order to emphasise the importance of the physical backgrounds of mechanics along with its mathematical aspects, the Congresses were initially named International Congresses of Applied Mechanics and only later renamed International Congresses of Theoretical and Applied Mechanics. In the twentieth century, mechanics has become a fully independent science in parallel with (applied) mathematics and physics.

In the 1950s, interesting monographs on the history of applied branches of continuum mechanics, namely the strength of materials, aerodynamics and hydraulics, were published. Two of them were written by great scientists of the twentieth century in fluid and solid mechanics – Theodor von Kármán and Stephen Timoshenko (1878–1972). Von Kármán became interested in the history of science in his old age, as is evidenced by his book *Aerodynamics: Selected Topics in the Light of their Historical Development* [1954] which covered mainly the development of the subject during the last hundred years and was closely connected with its author's own work in Western Europe and in the USA. In the field of the history of fluid mechanics, it is necessary to mention here also the interesting and comprehensive *History of Hydraulics* [1957] by Hunter Rouse (1906–1996) in partnership with his young collaborator Simon Ince, which presents a survey of the development of hydraulics from ancient to modern times.

In contrast with the first of the above-mentioned books, Timoshenko presented a fundamental treatise embracing the whole history of deformable solid mechanics with special attention to the strength of materials and structural mechanics. It is his *History of Strength of Materials, with a brief Account of the History of Theory of Elasticity and Theory of Structures* [1953], written on the basis of lectures on the history of strength of materials that he had given during twenty-five years to students in engineering mechanics in the United States. The book was compiled during study of a huge amount of material in western languages and in Russian, referring to more than 900 various sources – from Galileo to the time before World War II (and partially even afterwards).[8]

Timoshenko began his professional activity in Russia and continued it in the USA after the Russian upheaval of 1917 and is the greatest scientist and engineer of the twentieth century in this field. His *History* contains a deep analysis of the development of the Strength of Materials and Structural Mechanics including the first third of the twentieth century. The marvellous historical essays of Timoshenko on these developments are extremely informative and represent the first real milestone in the history of strength of materials and structural mechanics.

Almost simultaneously with Timoshenko's treatise, there appeared in Russia the *Essays in the History of Structural Mechanics* [1957] by Sergei A. Bernshtein (1901–1958). These covered a comparatively limited range of problems in four parts devoted, respectively, to the bending and buckling of rods (including problems of their design on the basis of their working and limit state), to the strength of arches and vaults, of trusses, and especially continuous-beam design.[9]

Among later essays on the history of mechanics, there should be mentioned *Geschichte der mechanischen Prinzipien und ihrer wichtigsten Anwendungen* von István Szabó (1906–1980) [1977] and, especially, *A History of Theory of Structures in the Nineteenth Century* by Thomas Malcolm Charlton [1982].

On a side note, it should be mentioned that historians of mechanics in general, and structural mechanics in particular, have received certain purely bibliographical assistance during the last decades by the Garland series of *Bibliographies on the History of Science and Technology*. Its volume devoted to Civil Engineering prepared by Darwin H. Stapleton [1986] should be mentioned specially. Among other, larger bibliographical aids, there is a list of publications on Fatigue (from 1838–1950) by John Y. Mann [1970].

The work of Clifford Truesdell

A new era in the history of mechanics was opened up in the middle of the 1950s by Clifford Truesdell (1918–2000) with his historical studies, including his basic investigations in the history of general principles and methods of

---

[8] The Russian translation of Timoshenko's *History* (1957) contains many additional references to the Russian literature from the 1920s and 1930s when Timoshenko worked in the USA.

[9] In particular, Bernshtein noted that the first design of continuous beams was done in 1808, before Navier, by Johann Eytelwein (1764-1849). He also referred to interesting Russian works by Aleksei Gvozdev (1897-1986) from the second quarter of the twentieth century.

rational continuum mechanics from the end of the seventeenth to the beginning of the nineteenth century, founded on a deep analysis of primary sources. For the first time, a young, creatively-active scientist in the field of rational mechanics was absorbed in a historic-critical analysis of the work of that time from the point of view of modern science.[10]

Truesdell received his initial serious interest in rational continuum mechanics in general, as well as in the history of mechanics, from his elder colleague Professor Paul Neményi (1895–1952), an immigrant from Hungary, with whom he had the luck to begin to work in the 1940s. While Neményi practically did not publish any essays in the history of mechanics, he influenced Truesdell by his extreme versatile interests and erudition. Neményi's sole short paper on the history of hydrodynamics was published posthumously by Truesdell [Neményi 1962]. As a concluding remark to this publication, Truesdell added,

> *That the history of fluid mechanics has been studied so little is one of the reasons for publishing the foregoing essay now. The fairly numerous recent works on mechanics as a whole do not add much to the material known to Neményi* [Neményi 1962: 86].

Thus influenced by Neményi, Truesdell began his fundamental historical investigations in connection with the preparation of Leonhard Euler's *Opera omnia*. To start with, he was invited to edit Euler's papers on fluid mechanics, constituting two volumes of the second series of the *Opera omnia* (vol.II-12, 13). This work fascinated Truesdell intensely; he studied all the primary sources and wrote two introductory articles of about 230 printed pages [Truesdell 1954, 1955], in which he gave a critical analysis of the entire development of pre-Lagrangean fluid mechanics and by far exceeded the contribution of Euler himself. His free command of several languages, including Latin, helped him in this enormous undertaking. Truesdell's next fundamental historical work consisted of comments on Euler's papers on flexible and elastic bodies published shortly before (1947–1957) by Fritz Stüssi (1901–1981), Henri Favre (1901–1966) and Ernst Trost (1911–1982), also as two volumes of *Opera omnia* (II-10, 11). Truesdell's introductory article: *The Rational Mechanics of Flexible or Elastic Bodies, 1638–1788*, formed a separate, 435-page volume of the *Opera omnia* [Truesdell 1960].

Truesdell's three introductory articles (altogether of about 660 pages) from Euler's *Opera omnia* presented, firstly and in all details, the critical history of the general principles and methods of rational continuum mechanics. By studying

---

[10] A thorough review of Truesdell's life and work has been published recently as a large, possibly too hagiographically styled monograph [Ignatieff and Willig 1999].

Euler's work, Truesdell not only became the greatest expert in the history of mechanics, but also an ardent admirer of Euler, whose astonishing genius and contribution to the development of mechanics always amazed him.

Some of Truesdell's next historical writings have been collected in a revised form in his *Essays in the History of Mechanics* [1968], translated also in other languages, and later on in the collection of his smaller reviews and articles with the challenging title *An Idiot's Fugitive Essays on Science* [1984].

Actually, Truesdell was the first to begin broad and serious investigations into the history of continuum mechanics within the frame work of the general history of mechanics and thus to extend immeasurably the range of the historical studies carried out up to that time. His work is distinguished by a profound knowledge of the source material and an original (and at times pungent) style of exposition.[11]

Moreover, his perfect, rich and colourful English should be referred to; it challenges the translator and casts its spell in all translations.

Landmark investigations by Truesdell in the history of mechanics (to which he added later also the history of thermodynamics [1980]) did not divert him from at least equally fundamental work in rational continuum mechanics, the modern format of which is to a significant degree due to Truesdell.

In Truesdell's obituary, published last year in co-authorship with me, my late friend Professor August A. Vakulenko (1925–2000) noted that the twentieth century had produced many famous scientists in the field of mechanics [Mikhailov and Vakulenko 2000]. Among the Western scientists of the first half of this century, he named Ludwig Prandtl (1875–1953), Theodor von Kármán, Richard von Mises, Sir Geoffrey Taylor (1886–1975), and Johannes Burgers (1895–1981). This list could be enlarged, but not too much. We have ventured to compare Truesdell's versatile and fundamental work during the second half of the twentieth century with that of Sir Geoffrey Taylor during its first half, in spite of the rather different styles and subjects of their investigations. Sir Geoffrey's distinguishing feature was a peculiar universalism; he obtained certain results of permanent significance in hydrodynamics, elasticity and plasticity theory, established the base of the theory of dislocations in crystals, performed some principal experiments, etc.

It should be emphasised that Truesdell was the only famous creatively-

---

[11] For example, Truesdell sharply criticised (and in this way shocked many of his contemporaries) Lagrange's *Mécanique analytique* for emasculating the rational content of Mechanics and distorting its history.

working scientist in rational mechanics who was also professionally engaged in the history of mechanics[12] and who can be rightly considered to be the initiator of the modern trends in the history of mechanics and, in particular, the founder of the general history of continuum mechanics.

In the 1950s, Truesdell came into close contact with the Springer Verlag – one of the greatest Western European publishing houses specialising in editing scientific literature. With the support of this publishing house, he founded and then edited – during many years – two fundamental journals: *Archive for Rational Mechanics and Analysis* (1957–) and *Archive for History of Exact Sciences* (1960–), which quickly acquired an extremely high reputation in their respective fields. Moreover, he established (1964) a series of *Springer Tracts in Natural Philosophy*, which published serious original monographs in mechanics, physics and applied mathematics. It is difficult to overestimate these of Truesdell's contributions to the development of the rational mechanics and history of mechanics; they reflect not only his creative, but also his organisational talents.

Apparently, Truesdell's influence is felt also in the volumes devoted to mechanics of the new edition of the *Handbuch der Physik*, which contain a number of scientific surveys with historical material on an unprecedented scale.

Clifford was not an easy man. His opinions were often very sharp and impartial and sometimes shocked his colleagues. As a really great man, he had many enthusiastic admirers and friends, among whom I dare to reckon myself, and also some foes, even among outstanding scientists who could not perceive and share his new approaches to Rational Mechanics.

The contribution of Edoardo Benvenuto

The development of the history of structural mechanics in the spirit of Truesdellian standards was realised by Edoardo Benvenuto (1940–1998) in his numerous studies and, particularly, in the two-volumes treatise *An Introduction to the History of Structural Mechanics* [1991] which has grown out of his large Italian monograph *La scienza delle costruzioni e il suo sviluppo storico* [1981]. Evaluating Benvenuto's treatise of 1991, Truesdell (who was very sparing with his praise) wrote in the Foreword,

This book is one of the finest I have ever read. To write a foreword

---

[12] Of course, Timoshenko was also professionally engaged in the history of mechanics, but he was rather considered to be a famous expert in applied mechanics than in rational mechanics. His engineering approach to the history of applied mechanics differs from Truesdell's analysis based mainly on pure physical and mathematical reasoning.

for it is an honour, difficult to accept... This book is the first to show how statics, strength of materials, and elasticity grew alongside existing architecture with its millennial traditions, its host of success, its ever-renewing styles, and its numerous problems of maintenance and repair [Benvenuto 1991: vii].

Truesdell emphasised here the originality of Benvenuto's presentation of the historical developments of structural mechanics characterised by linking scientific achievements with practical problems and progress of civil engineering and architecture. Such an approach shed new light on the development of structural mechanics, and Truesdell, being also an amateur and a fine expert of the arts, could not underestimate its merits.

Benvenuto's treatise is divided into two parts devoted to the early theories of the strength of materials and to the formation of the structural mechanics of arches, vaults and trusses during the eighteenth and nineteenth centuries. His analysis supplements substantially some sections of Timoshenko's treatise and is distinguished by its use of various Italian sources that are not sufficiently known in the total of Western European scientific literature.

In Italy, during the last decades, the history of science has developed very intensively and corrected many traditional points of view.[13] Truesdell has always revered the Italian culture, and was therefore especially inspired by the work of Benvenuto.[14]

I should say that by starting a new trend in the study of the development of mechanics, Edoardo Benvenuto has generated a new school in the history of mechanics, which was represented at its birth by a series of Symposia (from 1993 on) under the general title *Between Mechanics and Architecture* and by the

---

[13] The old Italians seem to have always been greatly concerned with a confirmation of their national contributions to science. They have often reprinted the works of their compatriots in various serial editions and separate collections (as, for example, *Raccolta d'autori italiani che trattano del moto dell'acque* in the eighteenth century). Nevertheless, the rich Italian scientific literature from the second half of the seventeenth to the beginning of the nineteenth century is insufficiently represented in the libraries of Western and Central Europe, whence it has been studied insufficiently until recent times by historians of science. A deep interest in the Italian science of former centuries has been revived in Italy during the last decades. Cf., in particular, the new series *Archivio della corrispondenza degli scienziati italiani* (1985 ff.) and *Biblioteca di «Nuncius» - Studi e testi* (1989 ff.), both published by Leo S. Olschki, Florence.

[14] In turn the Italian scientific community has appropriately recognised Truesdell's work. He obtained his first doctorate *honoris causa* from the Milan Polytechnic (1964), was awarded the Modesto Panetti prize and gold medal from the Accademia delle scienze of Turin (1967), Ordine del Cherubino from the University of Pisa (1978) and was a member of many Italian Academies, including the Accademia nazionale dei Lincei.

publication of the collection of articles under the same title, edited by him in co-operation with Professor Patricia Radelet-de Grave [1995]. I am sure that this interesting school will develop successfully and will take its rightful place in the history of mechanics.

Selected Bibliography

BENVENUTO, EDOARDO. 1981. *La scienza delle costruzioni e il suo sviluppo storico.* Florence: Sansoni. – xiv+915 pp.

———. 1991. *An Introduction to the History of Structural Mechanics.* Part 1: *Statics and Resistance of Solids.* Part 2: *Vaulted Structures and Elastic Systems.* New York, Berlin : Springer. – xxi+554 pp.

———. 1997. Adhémar-Jean-Claude Barré de Saint-Venant: The Man, the Scientist, the Engineer. *Atti dei Convegni Lincei,* n. 140: 7–34.

BERNSHTEIN, S.A. 1957. *Essays in the History of Structural Mechanics* [in Russian]. Moscow: Stroiizdat. [Repr. in S.A. Bernshtein, *Selected Works on Structural Mechanics,* Moscow: Stroiizdat, 1961: 272–448].

CHARLTON, THOMAS MALCOLM. 1982. *A History of Theory of Structures in the Nineteenth Century.* Cambridge: Cambridge University Press. – viii+194 pp.

CLAGETT, MARSHALL. 1959. *The Science of Mechanics in the Middle Ages.* Madison, WI: University of Wisconsin Press. –xxix+711 pp.

CLEBSCH, ALFRED. 1862. *Theorie der Elasticität der festen Körper.* Leipzig: Teubner. –xii+424 pp. [Cf. Todhunter and Pearson 1893, 2: 109–166.]

———. 1883. *Théorie de l'élasticité des corps solides de Clebsch. Traduite par MM. Barré de Saint-Venant et Flamant, avec des Notes étendues de M. de Saint-Venant.* Paris: Dunod. – xxi+900+32 pp. [Repr. New York: Johnson, 1966]. [Cf. Todhunger and Pearson 1893, 1: 199–286.]

DUGAS, RENÉ. 1950. *Histoire de la mécanique.* Paris: Dunod & Neuchâtel: Griffon. – 651 pp. [Repr. Paris: Gabay, 1996]. Engl. transl.: *A History of Mechanics.* Neuchâtel, 1955 [Repr. New York: Dover, 1988.]

———. 1954. *La mécanique au XVIIe siècle (des antécédents scolastiques à la pensée classique).* Paris: Dunod & Neuchâtel: Griffon. – 621 pp.

DÜHRING, EUGEN. 1873. *Kritische Geschichte der allgemeinen Principien der Mechanik.* Berlin: Grieben. – xxxi+513 pp. (3rd revised and enlarged ed., 1887, Leipzip: Fues [Rpt. Vaduz: Saendig, 1970].)

EYTELWEIN, JOHANN A. 1808. Handbuch der Statik fester Körper. Mit vorzüglicher Rücksicht auf ihre Anwendung in der Architektur. Bd. 3: Theorie derjenigen transcendenten krummen Linien, welche vorzüglich bei statischen Untersuchungen vorkommen. Berlin. – x+198 pp. (Cf. Abschnitt VII: Von der elastischen Linie.)

GIRARD, PIERRE SIMON. 1798. *Traité analytique de la résistance des solides, et des solides d'égale résistance, auquel on a joint une suite de nouvelles expériences sur la force, et l'élasticité spécifique des bois de chêne et de sapin.* Paris: Didot. – lv+238+48 pp. [There exists a German translation, 1803.]

IGNATIEFF, YU.A. and H. WILLIG. 1999. *Clifford Truesdell: Eine wissenschaftliche Biographie des Dichters, Mathematikers und Naturphilosophen.* Aachen: Shaker Verlag. – x+370 p.

HERIVEL, JOHN W. 1966. *The Background to Newton's* Principia*: A Study of Newton's Dynamical Researches in the Years 1664-84.* Oxford: Clarendon Press. – xvi+338 pp.

JOHNSON, WILLIAM. 1996. Isaac Todhunter (1820–1884): Textbook Writer, Scholar, Coach and Historian of Science. *International Journal of Mechanical Sciences*, vol. 38: 1231–1270.

KÁRMÁN, THEODOR VON. 1954. *Aerodynamics: Selected Topics in the Light of their Historical Development.* Ithaca NY: Cornell University Press. –ix+203 pp.

KLEIN, FELIX, and CONRAD MÜLLER, eds. 1907–1914. *Encyklopädie der mathematischen Wissenschaften mit Einschluss ihrer Anwendungen.* Bd.IV-4C. *Mechanik der deformierbaren Körper: Elastizität und Festigkeitslehre.* Leipzig: Teubner. – xv+770 pp.

LAHMEYER, WILHELM. 1843. Theorie der Kreisgewölbe (nach Petit bearbeitet). *Journal für die Baukunst*, vol.18: 207–254. (Cf. historical-bibliographical introduction, pp.207–210.)

MACH, ERNST. 1883. *Die Mechanik in ihrer Entwicklung historisch-kritisch dargestellt.* Liepzig: Brockhaus. – x+484 pp. (7th revised and enlarged ed., 1912) [There exist many later re-editions and English translations.]

MANN, JOHN Y. 1970. *Bibliography on the Fatigue of Materials, Components and Structures*, Vol.1. *1838–1950.* Oxford: Pergamon Press. – ix+316 pp.

MIKHAILOV, GLEB K. and AUGUST A. VAKULENKO. 2000. Clifford Truesdell and the Modern History of Mechanics [in Russian]. *Voprosy istorii estestvoznaniya i tekhniki* (Moscow), n. 3: 59–66.

MOODY, ERNEST A. and MARSHALL CLAGETT, EDS. 1952. *The medieval science of weights (Scientia de ponderibus).* Madison, WI: University of Wisconsin Press. – x+438 pp.

MONTUCLA, JEAN ETIENNE. 1802. *Histoire des mathématiques.* Nouv. éd. T.3. Paris: Agasse. (Cf. *Ch.X. De l'hydrodynamique. XI. Du cours des fleuves. XII. Des ondes et des oscillations des fluides*: 679–719.)

NAVIER, C.L. 1864. *Résumé des Leçons sur l'application de la mécanique à l'établissement des constructions et des machines.* 1éme section : De la resistance des corps solides. 3éme ed. avec des Notes et des Appendices par M. Barré de Saint-Venant. Paris: Dunod.

NEMÉNYI, PAUL. 1962. The Main Concepts and Ideas of Fluid Dynamics in Their Historical Development. *Archive for History of Exact Sciences*, vol. 2: 52–86.

PONCELET, JEAN VICTOR. 1852. Examen critique et historique des principales théories ou solutions concernant l'équilibre des voûtes. *Comptes Rendus Acad. sci. Paris*, t. 35: 494–502, 531–540, 577–587. [Cf. Todhunter and Pearson 1893, 2: 677–679.]

RADELET-DE GRAVE, PATRICIA and EDOARDO BENVENUTO, eds. 1995. *Entre mécanique et architecture/Between Mechanics and Architecture.* Basel: Birkhäuser. – 411 pp.

ROSENBERGER, FERDINAND. 1882–1890. *Die Geschichte der Physik in Grundzügen mit synchronistischen Tabellen der Mathematik, der Chemie und beschreibenden Naturwissenschaften sowie der allgemeinen Geschichte.* 3 Th. Braunschweig: F.Vieweg. – xii+175+407+827 pp.

ROUSE, HUNTER and SIMON INCE. 1957. *History of Hydraulics.* Iowa Inst. Hydraul. Res. – xii+269 pp. [Repr. New York: Dover, 1963.]

SAINT-VENANT, ADHEMAR BARRE DE. 1843. Note à joindre au Mémoire sur la dynamique des fluides, présenté le 14 avril 1834. *Comptes Rendus Acad. sci. Paris*, t. 17: 1240–1243.

———. 1864. *Historique abrégé des recherches sur la résistance et sur l'élasticité des corps solides.* Pp. xc-cccxi in Navier, *Résumé des Leçons ...* [1864]. [Cf. Todhunter and Pearson 1893, 1: 105–108.]

STAPLETON, DARWIN H. 1986. *The History of Civil Engineering since 1600: An Annotated Bibliography.* New York & London: Garland. – xxxiii+232 pp.

SZABÓ, ISTVÁN. 1977. *Geschichte der mechanischen Prinzipien und ihrer wichtigsten Anwendungen.* Basel: Birkhäuser. (2nd revised and enlarged ed., 1979; 3rd corrected and enlarged ed., 1987.)

TIMOSHENKO, STEPHEN P. 1953. *History of Strength of Materials with a Brief Account of the History of Theory of Elasticity and Theory of Structures.* New York: McGraw-Hill. – x+452 pp.

TODHUNTER, ISAAC. 1876. *William Whewell: An Account of His Writings with Selection from His Literary and Scientific Correspondence.* 2 vols. London: Macmillan. – xxxi+416 +439 pp. [Repr. as vols. 15 and 16 of William Whewell, *Collected Works,* S.l.: Thoemmes, 2001].

TODHUNTER, ISAAC and KARL PEARSON. 1886-1893. *A History of the Theory of Elasticity and the Strength of Materials from Galilei to Lord Kelvin.* Vol. 1 : *Galilei to Sant-Venant, 1639-1850* (1886). Vol. 2, pt. 1-2 : *Saint-Venant to Lord Kelvin* (1893). Cambridge: Cambridge University Press. – xvi+936, xiii+762 + 546 pp. [Repr. New York: Dover, 1960.]

TRUESDELL, CLIFFORD A. 1954. Editor's Introduction: Rational Fluid Mechanics, 1687–1765. *L. Eulerii Opera Omnia,* 2nd series, vol. XII. Zürich: Orell Füssli.

———. 1955. Editor's Introduction (The first three section of Euler's treatise on fluid mechanics (1766). The theory of aerial sound, 1687–1788. Rational fluid mechanics, 1765–1788). *L. Eulerii Opera Omnia,* 2nd series, vol. XIII. Zürich: Orell Füssli.

————. 1960. The Rational Mechanics of Flexible or Elastic Bodies, 1638–1788. Published as *L. Eulerii Opera Omnia,* 2nd series, vol. XI (2). Zürich: Orell Füssli.

————. 1968. *Essays in the History of Mechanics.* Berlin: Springer. – vi+384 pp.

————. 1980. *The Tragicomical History of Thermodynamics, 1822–1854.* New York: Springer-Verlag. – xii+372 pp.

————. 1984. *An Idiot's Fugitive Essays on Science: Methods, Criticism, Training, Circumstances.* New York, Berlin: Springer. – xvii+654 9+pp.

The Stradivarius violin, which used the kind of eighteenth-centry musical
wire used in the studies of William James Gravesande

Coping With Error in the History of Mechanics

Louis L. Bucciarelli[1]

My subject is error, failure, mis-step and/or the "incorrect" in the history of science, in particular in the history of mechanics. Focusing on the works of Clifford Truesdell, *The Rational Mechanics of Flexible or Elastic Bodies 1638-1788*, and of Edoardo Benvenuto, *An Introduction to the History of Structural Mechanics*, I explore how historians cope when confronted with developments crafted by the renowned that prove ill-fitted to what subsequently becomes the canonical form of theory, of concept, of ways of perceiving and analyzing the behavior of solids and structures. Taking error seriously moves the historian of science beyond the customary bounds on rational thought. Attempts at explaining the incorrect engenders speculation and conjecture about the influence of precedent and too strongly held ideologies – consideration of what an economist might call externalities or exogenous factors. It may lead to a fuller appraisal of empirical evidence and usually requires attempts to make reasonable patterns of thought now deemed irrational. Underlying my exploration is a tentative hypothesis – that the path to error, failure, mis-step and/or the incorrect in the history of science can not be, indeed is not, explained in the same way that we explain a correctly-formed concept, a successful extension of theory, or a competent experiment.

Setting the question

How do historians of science, in this instance two historians of mechanics, cope with the failings of their subjects? The question is problematic in more ways than one. Before even thinking of responding, one must answer a prior question: Error according to whom? Are we to judge against the standards, norms and beliefs of the past or do we take today's state of the art as our frame of reference? What do we allow as possible or probable cause of misdeeds? Do we limit our attention to instrumental reasoning and practices alone, as recorded in the texts produced by our subjects, or do we grant a role to what an economist would term 'externalities', e.g., social forces and/or institutional norms or intellectual currents. More fundamentally, how does one make reasonable an instance of irrational behavior on the part of one's historical actor – who otherwise produced solid, if not brilliant work? How to clarity confusion? Isn't there something contradictory at the root of the question itself?

---

[1] Professor of Engineering and Technology Studies, Massachusetts Institute of Technology, 77 Massachusetts Avenue, Cambridge, MA 02139 USA

This essay is not intended to resolve these questions. They provide an orientation, a sounding board, a backdrop to my main task, namely to compare and contrast how two historians of mechanics dealt with what they perceived as mis-steps of their subjects. We take, then, as criterion for error, a significant dissonance with the state of the art today, intentionally adopting an anachronical view in order to engage our authors, Clifford Truesdell and Edoardo Benvenuto, on their own terms. But first, we turn to what other historians have said on our subject.

Writing history when the past appears to make sense and the contributions of a Galileo or Euler resonate with our own is no light task in itself alone. What counts as scholarly research and writing in this case has been, and still is today, energetically debated. Joseph Agassi set two conditions on how the history of science should be written.

> The first maxim of enlightened or broad-minded historiography should be this: any interesting or stimulating story is good, and should count as history if it fulfils two conditions: (a) it does not often violate factual information easily accessible to its author, and (b) it does not present historical conjectures as if they were pieces of factual evidence [Agassi 1963: 74].

These strictures are meant to provoke more than enlighten. Good history is any interesting, stimulating story, not just a record of historical facts. Even apparent (easily accessible) facts can be violated, ignored, twisted around, if the historian judges the case otherwise. Conjecture is essential too, but be on guard: one must not mix historical fact with the historian's constructions. As such, Agassi's rules grant the historian considerable freedom to play with the facts – a prerequisite condition if one is to make history of mis-steps, misdeeds and failure.

While Agassi's 'positive views' – he calls them that – lay out what the historian should not do, Collingwood is more explicit in describing what is needed in order to make an interesting and stimulating story: interpolation, as well as critical assessment of the facts and source materials, is necessary. One must use one's imagination in fleshing out the past. But the imagination is to be employed in ways "...not ornamental but structural. Without it the historian would have no narrative to adorn."

> The imagination, that 'blind but indispensable faculty' without which, as Kant has shown, we could never perceive the world around us, is indispensable in the same way to history: it is this which, operating not capriciously as fancy but in its a priori form,

does the entire work of historical construction [Collingwood 1946: 241].

The sense one has here is that of necessity, not of possibilities. Just as through induction, the scientist develops a coherent theory that can be put to the test, the historian must employ his inductive powers imaginatively to fashion his story then confirm through additional facts and others' stories. Imaginative re-enactment of the past, the past thinking of history's agents, is Collingwood's way of doing history. The proper task of the historian is to penetrate "... to the thought of the agents whose acts they are studying."

There is a success-oriented flavor to Collingwood's proposal – in his Autobiography [1939] he explicitly rules out the possibility of explaining failure – a tilt toward the successful, productive, act of the past that leads Kragh to claim that Collingwood's method leaves us with the "...strange result that we can never say a scientist or philosopher concerned himself with problems that he could not solve; for we cannot have any historical knowledge of such problems" [Kragh 1987: 50]. This bias is apparent also in Agassi:

The only way we have to explain historical events satisfactorily is by the use of what has been called situational logic, by reconstructing the situation of historical people and their objectives, and by deducing from our assumptions the conclusion that their actual behavior was the most appropriate [Agassi 1963: 50].

We logically deduce and conclude from our assumptions that they behaved most appropriately. What, then, if they erred?

In all of this, both Agassi and Collingwood would agree that our reconstructions of past thought and deed should be in accord with the standards, beliefs and norms of the historical period we study. There is no contradiction here, only the challenge of keeping historian's conjecture apart from historical fact. A historian's responsibility is to re-enact using only the props and stagings of the past. We must in this continually struggle, in Agassi's words, to avoid "being wise after the event". He points to Koestler's recommendation that when approaching the past "...we see ourselves as children" – which Agassi, however, sees as insufficient.

A personal anecdote: As a first year student in engineering, I was mystified by the actions and words of my instructor, at the board, unfurling the fundamentals of the differential calculus. The symbols and words he deployed, the claims and transitions he made, his illustrations made no sense to me. I simply did not grasp the meaning and purpose of this display. Then one day I did understand; what was hash now had coherence. I achieved a new state of awareness; no longer a child, blind to the concept of derivative, I entered the game. But – and here is the relevant point – I can not now describe my prior state of not knowing. I have no picture of 'before' and to this day, though I have tried on several occasions to paint that picture, I can not do it. Was I thinking rationally but fooling myself? I do not know. I can only now say that then, I did not understand as I do now; the rest is a mystery.

The directive to understand the past in its own terms when our historical actor went astray has this flavor: We seek to understand the doings of our actors that strike us now, at first encounter, as mystifying. We strive to de-mystify, to decode, to clarify, to make rational that which pretends, in their terms, to reason but which to us is empty, incoherent, a jumble. Just as I cannot recover my beliefs and thinking prior to my enlightenment in coming to know the "true" meaning of the differential, once I experience its utility as a practitioner, I can not see without it.

Agassi is right; to approach the past as a child, to attempt to repress all that we know in our efforts to understand the thinking of predecessors will not work. We are right to put to use what we understand to be the case today, the concepts and principles of mechanics in their canonical form. To claim otherwise would be fooling ourselves. But here Agassi's second condition must be kept in mind: the need to distinguish conjecture from historical fact.

More contemporary scholars have directly addressed the question at hand here: how to cope with confusion and error. Lindholm is pessimistic

> *The new history of science in its most critical and adventurous branches still does not make room for the influence on scientists of confused or incoherent metaphysical ideas. Such influences are ignored or dismissed in the reconstructions because, I propose, it appears impossible to accommodate them without rendering the science a caricature. Yet, the resulting distortions and idealizations seriously inhibit our ability to understand science* [Lindholm 1981: 159].

Buchwald and Hong are more positive in setting out three criteria for evaluating missteps without falling prey to caricature or simply ignoring what may very well have been, as Benvenuto has argued, a significant event as measured by the contributions of others that followed.

> *This danger raises the question of what it means to assert that a scientist was mistaken. ... To justify doing so, the historian's critique should illuminate the point at issue in a historically-significant way; should not bring to bear knowledge that the subject could not possibly have possessed at the time; and should argue that*

the subject could reasonably have been convinced by a contemporary that he was in error [Buchwald & Hong: 20].

How did Benvenuto and Truesdell fare in this respect? To this we now turn.

General reflections

At a first reading, the similarities of Benvenuto's and Truesdell's histories come to the fore. Both center on the development of mechanics as an intellectual achievement in its own right. Their subject matter – whether theories for the construction of arches, explanations of the deflections large and small or of the fracture of beams, essays on the fundamental principles of elastic behavior – are always at center stage. Both apply today's state of the art in explaining the past.

Upon another reading, differences are apparent: Truesdell finds little of interest in explaining misdeeds. He is at times sympathetic to the missteps, the failings of the past; but then his tone of voice in these instances varies as a function of his esteem for the individual. More often he simply dismisses a text without giving more than a citation. Less frequently, he exhibits irony :

> *The reasoning [of Jordan de Nemore] is vague, qualitative, and insufficient if not erroneous, but the attempt at a precise argument to prove a concrete result in a domain never previously entered is of splendid daring.* [Truesdell 1960: 19]

Benvenuto is more gracious, never vicious, in his addressing of past failings. He allows that there may be something of interest in the confused contributions of those who don't measure up to Truesdell's standards. Commenting on de la Hire's graphic construction for determining the width of an abutment for a vault:

> *This "solution" is, at the least, a little too baroque. ... as we struggle amid a welter of lines, segments, circles and arcs, we find that we have not solved equation (10.5); it and the tangle of points and segments presented above give quite different results. Not surprisingly, this contradiction has never been clarified. Even less surprisingly, it gave rise to confusion and misunderstandings among de la Hire's critics and supporters. But de la Hire's theory has a certain historical and cultural interest. As so often happens, its value lay not in its results, which were seriously flawed, but in its stimulation of research and discussion* [Benvenuto 1990: 336].

Note in this, how we, the readers, are drawn into the game; "we find we have not solved the equation..." and we are implicitly invited in to resolve "...this contradiction [that] has never been clarified". Baroque theory, though seriously flawed, still is of interest to the historian. And it has value; it provokes further research and debate.

Benvenuto is not content to write off that which is dissonant with today's understanding. He presses and examines his sources, many of which are the same as Truesdell's, for different readings and ideas, even foreign or bizarre, that might explain why his actors (appropriately?) went astray.

For example, in summarizing Jacob Bernoulli's fundamental work (1706) on the curvature of elastic laminae, he lays out Bernoulli's three lemmas, which "..introduce the notion of a local stress-strain relationship" [Benvenuto 1990: 274]. The first two define what we would term linear elastic behavior, the third the possibility of non-linear behavior of the "homogeneous fibres" of the elastica. In this, Benvenuto quotes Truesdell's positive evaluation of Bernoulli's contribution, "It is the first time since Galileo's formula for rupture that a *material property* appears in rational mechanics" [Truesdell 1960: 106].

But when we come to the fourth, faulty, lemma, which "...asserts, in effect, that the moment required to bend a beam a given amount is independent of the position of the neutral axis...." [Truesdell 1960: 106], differences in treatment become apparent. Benvenuto, in the first instance, doesn't offer a modern reading but includes Bernoulli's text, including the accompanying figure (which Truesdell does not). He notes a precedent: Fr. Fabri, for one, promoted much the same "hoary thesis" in 1669. Retreating to inspect Fabri's thinking, we find Benvenuto advancing this provocative claim:

> ...for Hooke and Fabri, the proportionality between force and elongation was a universal property of bodies, reflecting the proportion which must exist between cause and effect. It is, in effect, more important as a measure of force than as a phenomenological law [Benvenuto 1990: 255].

The matter is not dropped there: he wonders, "Why was Bernoulli impelled to formulate a thesis that was not merely wrong, but superfluous as well?" Recalling the work Varignon who, just prior to Bernoulli's work, in a sense unified the work of Galileo and Marriotte on the behavior of beams – though there was a "dichotomy" twixt the two – Benvenuto conjectures that Varignon's tolerant treatment of these conflicting theories showed that the matter of what we would now call the location of the neutral axis, was not settled. Benvenuto goes on to note that while Parent disproved the Fabri-Bernoulli hypothesis in 1713, "... it took until the end of the eighteenth century to dissipate the old prejudice..." [Benvenuto 1990: 280].

Truesdell, on the other hand, will have none of this dallying about:

> Lemma IV asserts....., [This, as has been remarked many times, is false, and the two proofs BERNUOLLI presents are fallacious] [Truesdell 1960: 107].

Bernoulli is simply wrong, as many others have noted. He describes the two problems Bernoulli addressed, and in the second problem points out where he used the faulty Lemma IV in his final treatment of the bending of an elastic beam, but leaves it at that.

Truesdell, in short, sees no value in failure. Benvenuto speaks otherwise:

> Much as historians would like the history of ideas to move only onward and upward, the subject has a disquieting tendency to move upward, sideways, and occasionally downward. The importance of this crab-like motion should not be underrated. Often a partial failure may provide evidence – positive, as well as negative – which future research may use in the development of more fruitful theories [Benvenuto 1990: 404].

He is not afraid of reaching out to intellectual currents or philosophical trends apart from rational mechanics, to explain the thinking of his historical actors – much in the way Holton uses "thematics" to provide a basis for understanding the worthy deeds (but rarely misdeeds) of scientists. In his short but rich account of Galileo's achievements, Benvenuto sees a resonance in attempts by Bellarmino to reconcile the innovative thought of Galileo with Scripture with the nineteenth and twentieth century view of science found in the works of Duhem, Mach and Poincaré [Benvenuto 1990: 149].

Truesdell's references outside the bounds of rational mechanics on the other hand are few. But within his domain of expertise, he relentlessly pursues texts in order to make sure his claims. Still, he is selective: he has done all the work for the reader – done all the shopping, found all the bargains and specials and wrapped them in a provocative narrative; but it is narrative limited in scope, describing and valuing only the efforts of those historical actors who fixed the form of the rational mechanics of elastic bodies as we know it today.

Truesdell in particular

For an example of Truesdell's explanation of error, we turn to his evaluation of the work of Gravesande [1731], "a work to which writers of the eighteenth century occasionally referred", found in Part I, "Earliest Special Problems", of his essay.

... examination of his chapter On the laws of elasticity reveals it to be the report of a mass of ill conceived experiments garnished with bold assertions. He claims to establish the proportionality of deflection to load and length, but his experiment, employing specially designed and presumably precise apparatus, is imperfectly described and in any case would prove nothing at all [Truesdell 1960: 116].

Then follows a lengthy footnote in which Truesdell analyzes possible outcomes of Gravesande's experiment – outcomes dependent upon different assumptions Truesdell makes about the internal forces acting within, and the relative magnitude of deformations of, the test specimen – a wire stretched horizontally and loaded transversely at its midpoint.

He reports how Gravesande passed an elastic wire "...pulled taut by a *specified weight T*" over two wedges. The wire is stretched further by other weights hung from the center. Since the system is symmetric, Truesdell shows but half of the experiment in his figure, reproduced below. Gravesande finds that the deflections δ are proportional to the weights *P*. His footnote continues:

Since the forces exerted on the string by the wedges are not known, the problem is indeterminate.

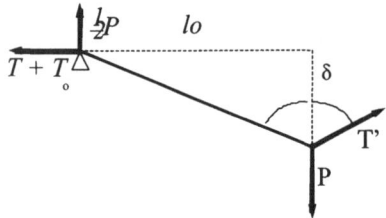

The most general possible system of forces acting on one half of the stretched string is shown in our sketch... where To is the unknown horizontal force exerted by the wedge. For equilibrium, we must have

$$P = 2T' \cdot \frac{\delta}{\sqrt{\delta^2 + l_o^2}}$$

$$\frac{\delta}{l_o} = \frac{\frac{1}{2} \cdot P}{T + T_o}$$

Thus if T is held constant, *and if* T_O *is constant or is much smaller than T, we must have* δ *[proportional to]* P, independently of any elastic law. *For this, no experiment is required* [Truesdell 1960:116].

So if Gravesande observed linear behavior, it might have been the case that T_O was constant or was much smaller than T. But in this case, the result obtains irrespective of the elastic behavior, independent of the form of the constitutive relation between the tension in the wire and its extension.

Truesdell considers another possibility – the particular case when the ends of the string are fixed:

We usually encounter another form of this problem, in which the end of the string is fixed. Then we have no concern with T or T_O, but, by Hooke's law,

$$T' = K \cdot \sqrt{\delta^2 + l_o^2} - L$$

where L, the initial length, may or may not equal l_o. Then (the equation above) gives

$$P = 2K\delta \cdot \left(1 - \frac{L/l_o}{\sqrt{1 + \delta^2/l_o^2}}\right) \cong 2K\delta \cdot \left(1 - \frac{L}{l_o} + \frac{1}{2} \cdot \frac{\delta^2 L}{l_o^3}\right)$$

Therefore δ *[proportional to]* P *holds for small deflections if and only if* L *[is not equal to]* l_o*. If* L = l_o *we get* P *[is proportional to the cube of]* δ *instead. This is a classic example to show that the response of a linearly elastic body may fail, for kinematical reasons, to be linear;*

Truesdell concludes Gravesande has proved nothing at all.

Thus Gravesande missed his chance twice over: had he set up the experiment properly, he would have failed to find the linear response he was looking for [Truesdell 1960: 117].

Truesdell frames the event in two ways: he obtains from equilibrium conditions alone a relationship between the transverse displacement of the wire at its midpoint and the load acting there. This relationship is linear if the horizontal component of the reaction force at the prism is constant or small with respect to the initial tension in the wire. The problem is indeterminate, so this force component remains an unknown.   No consideration of the deformation of the wire is required, though continuity holds presumably, and the relationship obtains, irrespective of the constitutive law, the relationship between force and elongation of the wire.

For example, the wire could be rigid, not stretched at all. In this case we must allow the wire to slide over the prism as it is loaded, so that the midpoint will deflect. Then equilibrium will linearly relate the load to the transverse displacement as long as at any state the horizontal force component at the prism is either the same in all cases or small with respect to the transverse load. That this component would remain the same is possible. That it would be small with respect to the initial tension is more likely, but in either case, I, and Truesdell, have nothing to support these conjectures if we rely upon an equilibrium condition alone. In an indeterminate problem such as this, there is a bounty of solutions from which to choose or to imagine. Indeed, one might conjecture that the additional force T_o acts the other way, i.e., is negative. With this as a possibility we might call upon the principle of sufficient reason and claim that T_o is zero. The point here is that if we constrain our reading to the mathematical meaning of this symbolic expression of equivalence, accepting only the denotative meaning of the symbols it contains, anything goes, so to speak. We need more of a narrative in order to extract its positive   implications. Equilibrium considerations alone tell us nothing about the force-deformation law of the wire.

Truesdell is correct in this.   But in claiming that this is possibly why Gravesande observed liner behavior, he is implicitly claiming something about Gravesande's experimental method and apparatus.

If the ends of the wire were truly fixed, then Truesdell observes Gravesande would have seen linear behavior if the original length of the wire L, in the state with no tension applied, were significantly different from the length l_o, the length of the wire with the initial tension applied but before deflected transversely by a load at midpoint ('significantly different' meaning that the ratio of the two lengths differed significantly from 1.0) Otherwise, the load varies non-linearly with the displacement.

But we can carry this analysis further and explore what the ratio of L to l_o might be for musical wire, the specimen Gravesande employed. Letting ε be the initial strain due to T, the initial tension in the wire, then, to the same order of approximation as Truesdell has made (i.e., assuming small deflections), we have $1 - L/l_o = \varepsilon$ and the above can be written

$$P \cong 2K\delta \cdot \varepsilon \left(1 + \frac{\Delta\varepsilon}{\varepsilon} \right)$$

where the second term within the bracket is the ratio of the additional strain in the wire, $\Delta\varepsilon$, produced by the transverse load P, to the initial strain.

Now if this additional strain is small (consistent with the assumption of small deflections) with respect to the initial strain, then Gravesande will indeed obtain a linear relationship between the lateral force and the transverse deflection. So it is not at all clear that Gravesande's work ought to be taken as "...a classic example to show that the response of a linearly elastic body may fail, for kinematical reasons, to be linear". In order to settle the case, we must dig still further. We turn to Gravesandes's "...mass of ill conceived experiments garnished with bold assertions".

Gravesande's apparatus is shown in Figure 1. The weight Π, attached to one end of the wire, hangs over the pulley and puts the wire in tension. (Note: to

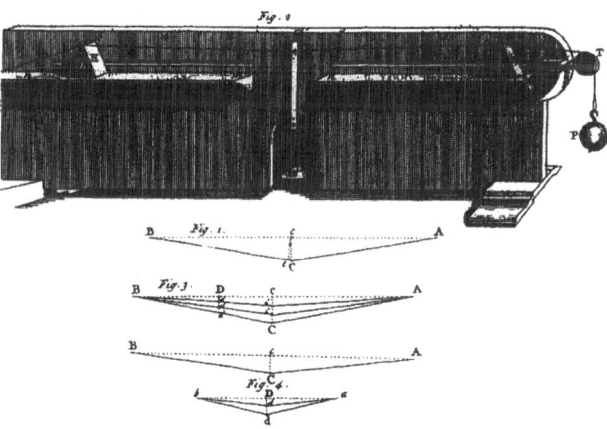

Figure 1. Plate 17, Figs. 1-4 from Gravesande, *Mathematical Elements of Natural Philosophy...*, 1731, p. 20

avoid confusion with the symbols used in Truesdell's critique, we replace Gravesande's P, his symbol for the weight suspended over the pulley, with the symbol Π). The prisms provide support at the ends of the active length of the wire; the length of the active length of the wire can be changed simply by changing the position of the prisms. He explains how the length of the wire is determined as the distance between the prisms

...for in the small inflexions made by hanging on the weights in C, concerning which alone experiments are made, the string is not

*moved upon the prisms, nor is the weight Π raised up, but only the
part AB is extended by these inflexions* [Gravesande 1731: 123].

Gravesande provides limited quantitative data in describing his apparatus and
specimens; he reports that with a wire length, an active length, of 2 ½ feet
between the prisms, the maximum midpoint deflection he observes as he loads
the wire transversely is ¼ in. As for the tensioning weight Π, he speaks of two
pounds, four pounds, six pounds [Gravesand 1731:123]; in another place, of 100
ounces [Gravesand 1731: 124].

What is missing, if our aim is to conjure up a likely ratio of the additional
strain to the initial strain, is a number for the cross-sectional area of the wire.
With this, we could estimate an initial stress due to the weight Π, hence the
initial strain. The additional strain is derived from the geometry of the deformed
state relative to the original state and we have information sufficient to that task
in the numbers cited above.

To proceed requires some imaginative, but grounded, conjecture. So we enter
the world of wires under tension of the first quarter of eighteenth-century
England. Gravesande gives us entry to that world, the world of stringed, musical
instruments:

> ... *we must consider Strings of musical instruments, and such as are
> of Metal; for Catgut Strings have a spiral Twist, and cannot be
> considered in the same manner as those fibres of which bodies are
> formed* [Gravesand 1731: 120].

Think, then, of his apparatus as a one-string musical instrument. The prisms
play the role of the bridge; the weight Π engenders an initial tension as one
might do in the tuning of a stringed instrument. The string itself is a metal wire
of the type used in a clavichord perhaps. Coulomb made use of the same type of
specimen in his experiments on the torsion of wires, some fifty years later
[Coloumb 1789].

Gravesande notes that elastic behavior, which he has defined in thoroughly
modern terms earlier on in volume 1, is not observed until some finite value of
the tension in the wire is obtained; this threshold he claims is unknown and yet
to be determined from experiment. But we conjecture that he added sufficient
weight at the end to produce a tone by the wire when plucked at mid span. From
this we can deduce a wire diameter.

For example: A weight Π of two pounds, a wire diameter on the order of 0.2
millimeters, with mass density 7.85×10^3 kg-m^3 spanning the 2 ½ ft (0.76 m)
between the prisms, would produce a tone in the neighborhood of 100 Hertz.
The initial strain in this case is on the order of .001. Now if the observed mid-

point deflection is on the order of 5 mm (e.g., ¼inch maximum) then the additional strain due to transverse loading would be on the order of .0001 The ratio of interest then is but .1, and this at the maximum transverse displacement. Thus Gravesande would have been entirely justified in claiming linear behavior. (Note: While the initial strain is significant, it would still remain well within the elastic limit for musical wire of the eighteenth century.)

But does his observation necessarily imply that the force-deformation relationship for a wire in tension is linear? No. For this we need to consider the possibility of such non-linearity and see where that leads, still respecting the situation as Gravesande described it – in particular the smallness of deflections relative to the active length of the string.

We return to the requirement for equilibrium, the first equation above, set $T = T + \Delta T$, where the second term is the additional tension due to the inflexion, and ignore terms of order $(\delta/l_0)^2$ with respect to 1.0, and obtain

$$P \cong 2T \cdot \left(\frac{\delta}{l_O}\right)\left(1 + \frac{\Delta T}{T}\right)$$

At this point we have said nothing about the nature of the force-deformation relationship for a wire.

Now, if the behavior is linear, we have our earlier result since $\Delta T/T = \Delta\varepsilon/\varepsilon$ in this case. If, on the other hand, the behavior is non-linear and the ratio of additional tension due to inflexion to the initial tension is small – certainly a mathematical possibility – we would still observe the deflection increase linearly with the transverse load P. This would seem to justify Truesdell's claim that Gravesande has proved nothing.

But, while analytically possible, we can ask if, practically speaking, this remains a possibility. If deflections are small, would one be able to observe and measure the non-linearity? It is true, as Truesdell has noted, that the response of a linearly elastic body may fail, for kinematic reasons, to be linear. But the situation here is the other way around. The macro response is linear; the question is, does this entail linear elastic behavior at the micro level or, alternatively, from a practical perspective, how can one determine the form of a non-linear relationship at the micro level from an experiment on real materials of the sort our historical actors considered and had available in which the macro behavior is linear? If the macro behavior is non-linear, we can not deduce that the micro behavior is non-linear. If the macro behavior is linear, what experiment would detect and define the form of the non-linear constitutive relation?

Turning to Gravesande's argument, we find the following analysis linking the linear behavior he observed to linear elastic behavior at the micro level. Referring to 'Fig. 3' in Figure 1,

> *Let the wire AB be so inflected as to acquire the positions AcB, AcB, and ACB, set so that in the greatest inflexion the sagitta (the downward displacement) may not be 1/4 inch long, supposing the wire 2 feet and a half; In those cases the lengthenings of the string are very small, therefore they are in the ratio of the forces that produce them, and they serve to express them; let cD express the force by which a string is stretched when it is not inflected, and with the center B describe the circle Dd; the lines dc, dc, dC, which are longter than cD by the quantity by which the fibre was lengthened in every case, express the whole forces, by which the fibre is stretched in every case. But there the arc Dd is hardly of one degree, and D is always far enought distant from the Point c, wherefore Dd may be looked upons as a right line parallel to cC, and the lines cd, cd, Cd have the same ratio to the lines cB, cB, CB. Therefore the point C is always drawn towards Bc and A, by Forces proportionalble to the line CB or CA; and the force by which the wire is inflected, whose direction is along cC, is as the double sagitta, or as the sagitta itself.* Therefore in all the least inflexions of a chord, musical string or wire, the sagitta is increased and diminished in the same ratio as the force with which the chord is inflected [Gravesande 1731: 124].

Gravesande has proved the following: Assuming the string behaves linearly, i.e., "...In those cases the lengthenings of the string are very small, therefore they are in the ratio of the forces that produce them and they serve to express them..."; then if the deflections are small, he proves, through geometrical reasoning (similar triangles), that the total tensions in the string, "the whole forces", are proportional to the lengths of the string in the deflected state. Now the component of the total force in the string in the vertical direction is proportional to the ratio of the vertical deflection to the length of the string in the deflected state. Hence the force resultant of the two tensions in the vertical direction, "the force by which the wire is inflected, whose direction is along cC", is proportional to the vertical deflection (or twice the deflection).

In short, Gravesande proves that if we assume at the outset linear elastic behavior of the body, and if deflections are small, then macro behavior is linear. This is something but not everything, as we have noted.

His awareness of the complexities of the determination of the constitutive behavior of strings lies outside of Truesdell's field of view. Gravesande sees his experiment as a way of tying the macroscopic behavior of real solids and structures to their microscopic constitution and behavior – to reveal what is going on at a finer scale which would provide a basis for understanding the elastic behavior of solids other than musical strings.

> *... what this proportion is (between the force and extension of a wire) must be determined by experiment on metal. But these wires are scarce sensibly lengthened, the Proportions of Lengthening cannot be directly measured; therefore they must be measured by another Method* [Gravesande 1731: 121].

Today the disjunction between the micro and macro behavior is hardly felt. And to sense the nature of the problem as it was felt in the eighteenth century is perhaps out of our reach, just as I can not picture my lack of understanding of the differential operator back ages ago, in my first year at university. In our development of a theory of elasticity we cite Hooke's Law without thinking there might be have been any disjunction between what Hooke observed with his springs and beams and the constitution of solids at a micro level. But to the eighteenth century savant who saw in materials particles and corpuscles attracting and repulsing one another according to unknown laws, our notion of continuum would have been seen as a metaphysical fantasy. A continuum, after all is a construction, a mathematical idealization, a way of modeling – a very powerful (Cauchy showed how powerful) and effective way of modeling – but it is an appropriate model only at a certain scale, a certain resolution.

Gravesande knew the distinction between a model and – shall we say – the real constitution of bodies. He asserts:

> *All bodies, in which we observe Elasticity, consist of small Threads or Filaments,* <u>*or at least may be conceived as consisting of such Threads*</u>*; and it may be supposed that those Threads laid together make up the Body; therefore that we may examine Elasticity in the Case which is the least complex, we must consider Strings of musical Instruments...* [Gravesande 1731: 120, emphasis mine].

The representation of a solid as a collection of fibres may be taken as a model, a convenient way of representing an elastic solid that avoids claims about the real constitution of matter and one for which he has an elementary and fundamental sample – the string of a musical instrument.

Independent of these suppositions, we must allow that Gravesande's experiment is worthy of more than the too-hasty and incomplete treatment that Truesdell accords it. Indeed, there is still more to be done.

Truesdell questions Gravesande's competence as an experimentalist: he refers to "…(his) mass of ill conceived experiments …. imperfectly described" though we may presume his apparatus was precise. Should we presume such? Did Gravesande set up his experiment improperly? There is but one way to find out, namely by attempting to replicate what Gravesande did. While rational analysis can go a long way in the evaluation of experiment as well as theory – particularly if we allow, as we have here, the analyst to draw upon his or her own experience with solids and structures both in theory and practice; if we are serious about trying to understand the past, including its misfits, in its own terms, then there is no better way to force one out of today's secure understandings and to avoid being "wise after the event" than setting oneself the task of replication.

This is no easy task if one insists, in accord with the norms of good, scholarly, historical research, that one use only the materials, fabrication techniques, and instrumentation that were available and employed at the time of the event. Just as we can question whether it is possible to reconstruct rationally the thinking and theory constructions of our historical actors employing only the ideas, concepts, beliefs and norms of the historical period, so too redoing an experiment according to the same rules may well be impossible. And this is not solely because their techniques are not sufficiently described in texts, as Truesdell observes and as we have noted, but rather as much because critical facets of infrastructure – craft technique for example – are no longer available to us. Just as we may be mystified by their thinking, so too we can remain in the dark about how they managed to make their apparatus work as they reported. One learns in this way, if nothing else, how difficult it is to repeat the mistakes of the past.

Conclusion

We began this essay with a question: How do historians of science cope with error? We end with two more fundamental queries: Why study error at all? Is its explanation possible?

I can think of at least three reasons why it is important to study the failings of the past. First, it's an intellectual challenge, one in some ways more rewarding if accomplished than that of explaining how a successful contribution of a Euler or Galileo prefigured today's state of the art. Second, it brings to the fore the question of what resources, what knowledge and know-how, we may legitimately bring to the table in our historical explanations. A successful development, one which resonates with today's state of the art, explains itself in a sense. Finally, there is Benvenuto's point: We should try to account for failings and error because, while wrong, confused and/or faulty, that which draws our attention may have had significant impact on future work.

Is it possible? There are three points to make here: First, as our analysis of Truesdell's critique of Gravesande's work shows, what might at first perusal be dismissed as error, upon a closer reading, may not be seen as error at all. Explanation in this case is possible; what was erroneous is no more. Second, we may be able to satisfy the three criteria of Buchwald and Hong and conclude that, indeed, our historical actor was simply wrong as judged within the context of the times. The counterfactual character of their third criterion, however, is problematic. (As Benvenuto might say at this point, this question is left for another time.) Finally it may not *be* possible: in this case we throw up our hands and allow that our historical actor was truly confused. We can not construct a contemporary who might have straightened out our misguided savant. All would be, and perhaps were, similarly confused. Perhaps we can do no more in this case than look forward, following Benvenuto's advice, and seek out, in the more successful work of others, citations to that which mystifies us, but evidently had meaning transcending the author's text. But that too is a question to leave for another time.

References

AGASSI, J. 1963. *Towards an Historiography of Science*. Gravenhage: Mouton & Co.

BENVENUTO, E. 1990. *An Introduction to the History of Structural Mechanics*. Springer-Verlag.

BUCHWALD, J.Z. and S. HONG. Forthcoming. Theory, Experiment, and Practice in Nineteenth-Century Physics: Historiographical Issues. In *From Natural Philosophy to the Sciences: Historiographical Essays on Nineteenth-Century Science*, D. Cahan, ed. Chicago: University of Chicago Press.

COLLINGWOOD, R.G. 1939. *An Autobiography*. London and New York: Oxford University Press.

————. 1946. *The Idea of History*. Oxford University Press.

COULOMB, C-A. 1784. Recherches théorique et experimentales sur la force de torsion et sur l'élasticité des fils de metal. *Mem. De l'Acad. Roy. Des Sci.*

GRAVESANDE, W.J. 1731. *Mathematical Elements of Natural Philosophy Confirmed by Experiments, or an Introduction to Sir Isaac Newton's Philosophy*. Written in Latin, by William James Gravesande...Translated into English by J.T. Desaguliers..., vol. 1, 4th ed. London.

KRAGH, H. 1987. *An Introduction to the Historiography of Science*. Cambridge University Press.

LINDHOLM, L.M. 1981. Is Realistic History of Science Possible? In *Scientific Philosophy Today: Essays in Honor of Mario Bunge*. J. Agassi and Robert Cohen, eds. Dordrecht: D Reidel.

TRUESDELL, C. 1960. *The Rational Mechanics of Flexible or Elastic Bodies, 1638-1788*. Introduction to *Leonhardi Euleri Opera Omnia*, 2nd series, vol. XI, Zürich: Orell Füssli.

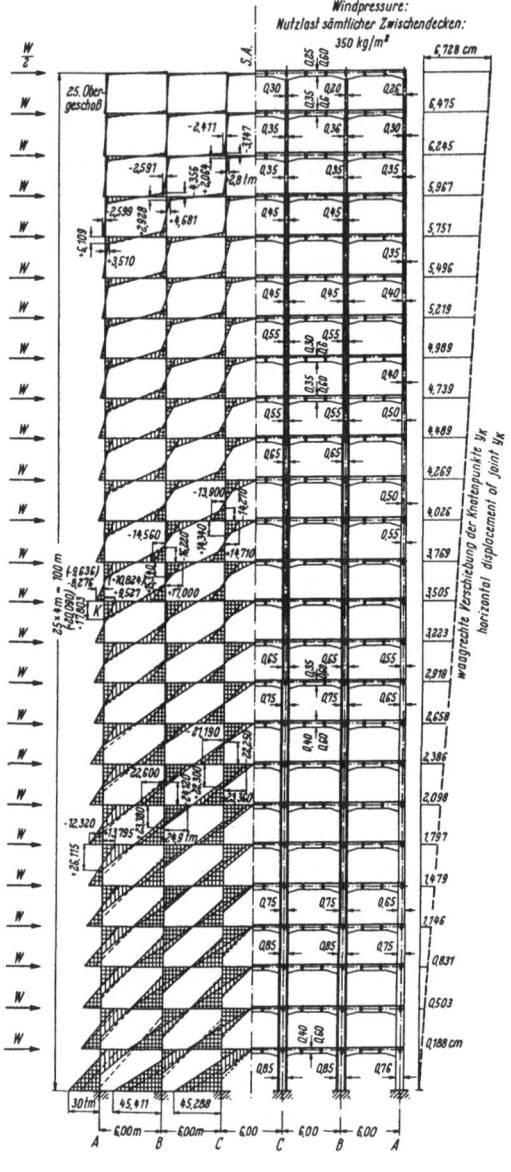

waagrechte Verschiebung der Knotenpunkte y_k
horizontal displacement of joint y_k

THE DEVELOPMENT OF THE DEFORMATION METHOD

Karl-Eugen Kurrer[1]

A hundred years ago, the force method formed the core of classical structural calculations. Today, the deformation method is one of the most important pillars of modern structural mechanics. The deformation method played a decisive role during the transition from classical structural calculations to modern structural mechanics in the 1950s and 1960s. The internal structure of this method was particularly suitable for implementation on computers, and today it is also used as an introduction into the basic principles of modern structural mechanics (for instance the Finite Element Method). The development of the deformation method clearly shows the dialectic interaction between the logical and the historical. Goethe's proposition that the history of science is science itself was also a central theme for Edoardo Benvenuto. This contribution shows that the deformation method also has a genuine techno-scientific origin in its application to the problem of secondary stresses in riveted steel framework structures.

Introduction

A hundred years ago, the force method formed the core of classical structural theory. Today, the method of deformations is one of the most important pillars of modern structural mechanics. The method of deformations played a decisive role during the transition from classical structural theory to modern structural mechanics in the 1950s and 1960s. The internal structure of this method was particularly suitable for implementation on computers, and today it is also used as an introduction into the basic principles of modern structural mechanics. The development of the method of deformations clearly shows the dialectic interaction between the logical and the historical.

Goethe's proposition that the history of science is science itself was also a central theme for Edoardo Benvenuto, as indicated by the title of his book *La Scienza delle Costruzioni e il suo sviluppo storico*. Benvenuto was the first person to demonstrate that Alfred Clebsch, in his book *Theorie der Elasticität fester Körper* published in 1862, had already developed the basic concept of the method of deformations.

This article shows that the method of deformations also has a genuine techno-scientific origin in its application to the problem of secondary stresses in riveted steel trussed frameworks. The theory of secondary stresses in trussed frameworks developed by Manderla, Winkler, Engesser, Ritter, Landsberg, Müller-Breslau

[1] Ernst & Sohn Verlag, Editor-in-Chief, *Stahlbau*, Bühringstraße, 10 - 13086 Berlin – GERMANY

and others in the 1880s was relevant for structural theory, because it discussed the basic ideas of the second-order theory and introduced displacement values as unknowns for determining the forces within statically indeterminate trussed frameworks. Otto Mohr's contribution from 1892-93 is not only the historical-logical keystone of the theory of secondary stresses in trussed frameworks, but also the cornerstone for the development of the method of deformations. But it was only the large-scale introduction of reinforced concrete after the year 1900 that required a clear, efficient and closed calculation theory for statically-indeterminate frames commensurate with the monolithic character of this construction method. In 1914, the Danish engineer Bendixsen picked up this clear and yet so simple procedure from Mohr and applied it not only to trussed frameworks with stiff joints, but also to braced and non-braced frames. In 1921, Ostenfeld brought the conditional equations for displacements into the same form as the elasticity equations for the force method, which were well known to structural engineers; he introduced the term "method of deformations" and recognised its formal duality with the force method. In 1927, Ludwig Mann worked through the method of deformations against the background of Lagrange's *Mécanique analytique*. Up to the 1950s, reflection on the duality of the force method and the method of deformations remained a favourite subject of structural research. It was only the application of matrix calculations that enabled aircraft construction researchers to integrate structural theory within computer mechanics and modern structural mechanics. This was also the beginning of the substitution of the force method by the method of deformations.

The contribution of the mathematical elasticity theory

Towards the end of the nineteenth century, the mathematical elasticity theory was well established. In the first edition of his book *Treatise on the theory of elasticity*, published in 1892 and 1893, August Edward Hough Love (1863-1940), provides a complete overview of this scientific discipline. The second edition of the book was translated into German by Aloys Timpe at the suggestion of Felix Klein (1849-1925). Timpe's translation, published in 1907 under the title *Lehrbuch der Elastizität* [Love 1907], starts with a thirty-eight-page historic introduction describing the significant stages of the development of the mathematical elasticity theory and culminates in the formulation of its disciplinary identity:

> *The history of the mathematical theory clearly shows that the development of the theory was not exclusively guided by its suitability for technical mechanics. Most men, on whose research it was based and developed, were more concerned about scientific*

than about material progress, more interested in understanding the world than making it more comfortable [Love 1907: 37].

In particular, Love notes that the solution methods of the mathematical elasticity theory form an important part of analytical theory, "which has a high significance in pure mathematics" [Love 1907: 37]. Even for the analysis of more technical problems it is suggested that,

> *people were generally more interested in the theoretical side than the practical side of these questions. Gaining insight, Love continued, into what happens during impact, bringing the theory of the behaviour of thin bars into harmony with the basic equations – for most of the men who can take credit for the elasticity theory these and similar aims were more enticing than striving to find means for savings in machine construction or to determine safety conditions in buildings* [Love 1907: 38].

Love's men of mathematical elasticity theory of the nineteenth century were more interested in causality than in finality. They moved more in the ideal world of mathematical objects than in the real world of engineering objects. They saw themselves more as discoverers of laws of nature than as inventors of technical artefacts, and saw their discipline more as theoretical natural science than as a basic practical discipline of classic engineering science. They interpreted their discipline more in the context of the contemporary philosophical discourse than in the context of the material needs of industrialisation, and saw themselves more as men of university science than as men of technological action.

What was the contribution of the mathematical elasticity theory of the nineteenth century to the method of deformations?

Elimination of stresses or of displacements, that is the question

As is generally known, the equilibrium conditions, the law of materials and the kinematic relationships lead to fifteen equations or partial differential equations with fifteen unknown scalar functions of three variables, namely,

- three displacements;

- six strains;

- six stresses.

This triadic structure characterises the logical core of the elasticity theory. In principle, the elasticity theory offers two paths for finding a solution: elimination of stresses and elimination of displacements.

If, for the case of complete linearity and for homogeneous and isotropic bodies, the stresses and strains are eliminated from the system of equations, a vectorial differential equation remains

$$\Delta w + [1/(1 - 2v)]\text{grad}(\text{div}w) = [2(1 + v)/E]k. \tag{1}$$

These three coupled partial differential equations for the calculation of the displacement vector w from the volume forces k and the two material constants E (modulus of elasticity) and v (Poisson's ratio) and the geometric boundary conditions were named after Gabriel Lamé (1795-1870) and Claude Louis Marie Henri Navier (1785-1836). One could subsume approaches leading to solutions of the Lamé-Navier displacement differential equations under the term "method of deformations" of the mathematical elasticity theory.

The second route is via elimination of displacements and strains – again for the case of complete linearity and for homogeneous and isotropic bodies – and leads to the tensorial differential equation named after Eugenio Beltrami (1835-1900) and John Henry Michell (1863-1940)

$$\Delta S + [1/(1 + v)]\text{grad}(\text{grad}\sigma) = -\{\text{grad}k + \text{grad}^T k + [v/(1 - v)](\text{div}k)\,I\}. \tag{2}$$

Taking into account the dynamic boundary conditions, the components of the stress tensor S (σ diagonal sum of the stress tensor, I unit tensor) can be calculated from these six coupled partial differential equations. Approaches leading to solutions of the Beltrami-Michell stress differential equations could be termed force method of the mathematical elasticity theory.

In the literature, the first route via the Lamé-Navier displacement differential equations and the geometric boundary conditions (specification of the displacements on the surface of the body) is called the first boundary value problem, and the second route via the Beltrami-Michell stress differential equations and the dynamic boundary conditions (specification of the forces on the surface of the body) is called the second boundary value problem of elasticity theory [Leipholz 1968: 116-121]. As is so often the case, the elasticity theory offers yet another, third route: The third boundary value problem describes the case where forces are specified for part the surface of the body, and displacements are specified for another part [Leipholz 1968: 121].

Having thus provided a brief logical description of the deformation and force methods within the context of spatial elasticity theory, we will now move on to a historical/logical description of the method of deformations in the context of elastic truss systems.

THEORIE

DER

ELASTICITÄT FESTER KÖRPER

VON

Dr. A. CLEBSCH,

PROFESSOR AN DER POLYTECHNISCHEN SCHULE ZU CARLSRUHE.

LEIPZIG,

DRUCK UND VERLAG VON B. G. TEUBNER.

1862.

Figure 1. Title page from the first German-language monograph about the elasticity theory [Clebsch 1862]

An element from the ideal object world of the mathematical elasticity theory: the elastic truss systems

With his *Theorie der Elasticität fester Körper* (*Theory of elasticity of solid bodies*) [Clebsch 1862] (Figure 1), published in 1862, Alfred Clebsch (1833-1872), who taught mathematics at the Polytechnic School in Karlsruhe, wanted to provide a textbook of elasticity theory, "which (...) was intended to fully cover the theory and practice of this discipline..." [Clebsch 1862: V]. Having developed the basic equations of spatial elasticity theory and substantiated them with examples over 189 pages, in the 166-page strong comprehensive second part Clebsch elaborates the basic equations for bars and thin plates. In the 69-page third part, Clebsch applies the equations derived in the second part to elastic truss systems; on pages 409 to 420, we find Clebsch's method of deformations, which was later analysed by Edoardo Benvenuto (1940-1998) [Benvenuto 1991: 492-498]. Clebsch describes the basic idea of the method of deformations for calculating the displacements of the joints in elastic truss systems as follows:

> *Here too, the general principle will be to initially treat the displacements of the joints as known parameters, to determine from them the elastic forces with which the bars react in their joints, and finally to establish the equilibrium conditions for the external and elastic forces acting in the joints; these equations will then enable the introduced displacements to be calculated* [Clebsch 1862: 413].

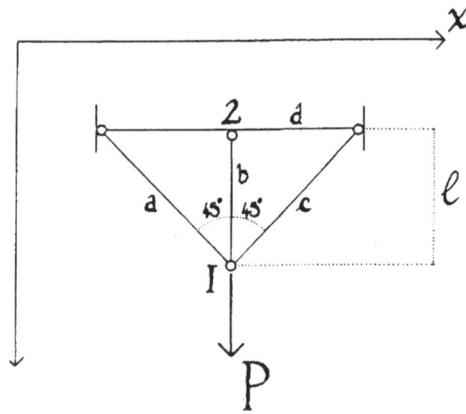

Figure 2. Analysis of an elastic truss system after Clebsch [Benvenuto 1991: 497]

Clebsch developed the method of deformations analytically, i.e., without graphic representation. Benvenuto can take credit for discovering Clebsch's method of deformations for the scientific discipline of the history of structural mechanics and presenting it in a comprehensible manner. Clebsch explained his method of deformations using the example of the calculation of the joint displacement in a simple hinged trussed framework (without bending) and generalises it for elastic truss systems. Figure 2 shows the 3-hinge system investigated by Clebsch with the vertical tension bar 1-2 subjected to load P in joint 1. The unknown quantities are the displacements of joints 1 w_1 and 2 w_2 in direction y. In the first step, Clebsch express the forces in bars a, b and c as a function of the displacements w_1 and w_2; these displacements are then determined from the equilibrium conditions for joints 1 and 2.

Clebsch clearly recognised that the joint equilibrium conditions form a system of linear equations for determining the joint displacement. "The problem," he wrote, "is therefore reduced to the solution a system of linear equations, and can thus be regarded as fully solved" [Clebsch 1862: 418]. The fact that the method of deformations for the analysis of elastic truss systems formulated by Clebsch was not adapted by engineers can be explained as follows: at the time, trussed framework theory was more interested in member forces than in joint displacements; it was much easier to investigate statically determinate hinged trussed frameworks using the methods of graphic static that emerged during the 1860s; Clebsch's monograph did not meet the clarity criterion for engineering textbooks.

Clebsch's textbook was thus mainly used by men whose cognitive objects represented elements of the ideal object world of the mathematical elasticity theory. Nevertheless, with his method of deformations Clebsch anticipated the mathematical methodology of theory formation for a techno-scientific fundamental discipline based on elastic truss systems: the structural theory.

From hinged trussed frameworks to trussed frameworks with rigid joints

Structural theory was developed between the 1820s and the 1890s. This period can be sub-divided into three phases. The <u>constitutional phase</u> is characterised by the formulation of the programme of structural theory and the completion of the technical bending theory of elastic beams by Navier. The centre point of the <u>establishment phase</u>, which followed in the late 1840s, was the creation of the trussed framework theory by Dimitrij Iwanowitsch Jourawski (1821-1891), Squire Whipple (1804-1888), Karl Culmann (1821-1881) and Johann Wilhelm Schwedler (1823-1894). During the mid-1860s, the theory of statically-indeterminate trussed frameworks began to emerge, which was combined between the late 1870s and the 1890s with the technical bending theory to form the theory of linear-elastic truss systems. The force method developed by Heinrich Müller-Breslau (1851-1925) from the theory of statically indeterminate truss systems forms the keystone of the <u>completion phase</u> of this structural theory formation period.

The centerpiece of the structural theory formation period was undoubtedly the trussed framework theory. As load-bearing structures, trussed frameworks dominated civil engineering in the second half of the 19th century. Whilst numerous trussed frameworks had been constructed from timber in the past, the trussed framework construction method only really took off when iron materials began to be used as structural elements: In the 1840s and early 1850s, trussed frameworks were dominated by mixed systems made of timber, cast iron and wrought iron. With reference to such trussed frameworks, in the first of his two travel reports Culmann introduced the term *Fachwerk (trussed framework)* in 1851 and developed a trussed framework theory [Culmann 1851], implicitly assuming frictionless hinges at the joints. In the same year, Schwedler managed to logically differentiate between the material reality of trussed frameworks and their structural system – the hinged trussed framework (Figure 3):

> *Whilst the frames are thought to be of rigid construction, the small resistances caused by the small elastic bending at points a, d, c etc. are negligible compared with the resistance of the strut, or, which is the same thing, the individual framework components can be assumed to have a hinged joint in points a, d, c, etc.* [Schwedler 1851: 168].

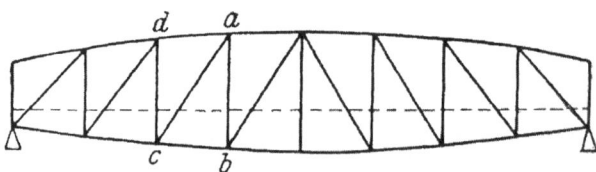

Figure 3. Schwedler's hinged trussed framework model [Schwedler 1851: 168]

For the first time, Schwedler thus accomplished the abstraction process that typifies structural theory from the physical load-bearing structure (actual trussed framework, e.g., timber framework) via the abstract load-bearing structure (model of the trussed framework or Culmann's *Fachwerk* notion) to the structural system (hinged trussed framework), i.e., the abstract load-bearing structure described through geometric/material properties for the purposes of quantitative examination. The invention of the structural system in the shape of the hinged trussed framework became the guiding concept for the development of the structural theory in the second half of the nineteenth century.

a)

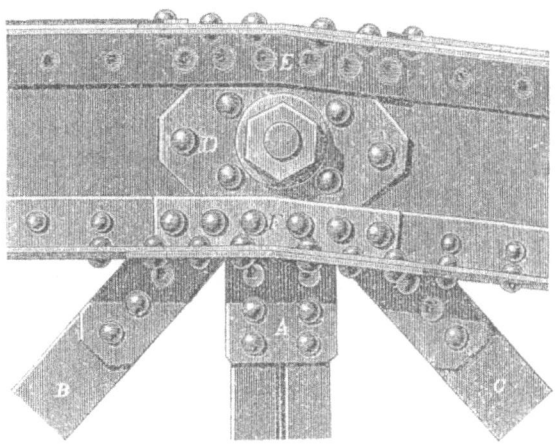

b)

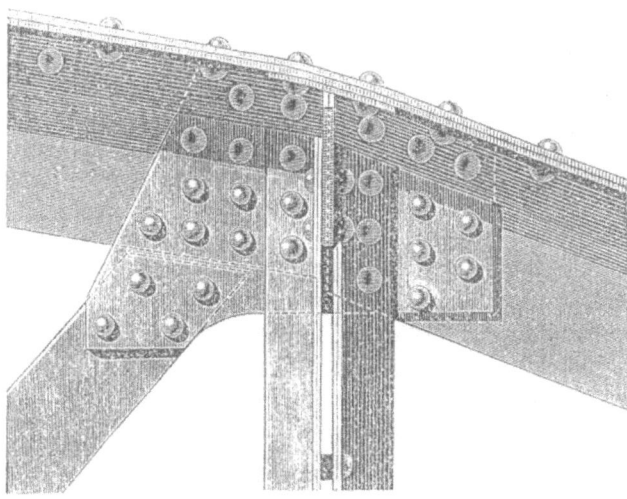

Figure 4 a) Hinged top chord joint of the bridge across the Brahe at Czersk, built in 1861 [Winkler 1872: 135]; b) Rigid top chord joint of the bridge across the Brahe near Bromberg, built in 1870 [Winkler 1872: 130]

An element from the real object world of the engineer: the iron trussed framework with riveted joints

The hinged trussed framework model of the trussed framework theory cannot conceal its proximity to the design language of machine construction. Werner Lorenz investigated the design thinking of August Borsig (1804-1854), who was successful both as a mechanical engineer and as a structural engineer (see [Lorenz 1990: 5] and [Lorenz 1997: 293-294]) and was able to show that Borsig regarded his buildings as machines [Lorenz 1990: 5]. Structural civil engineering had not yet quite completed its separation from the disciplines of machine and railway construction, nor from the shipbuilding and industrial mining metallurgy that were so typical for the industrial revolution. "The early history of structural civil engineering," Wieland Ramm stated recently, "is largely identical with the history of iron construction" [Ramm 2001: 640]. Iron construction as the core of structural civil engineering began to form the basis of an independent design language as early as the 1850s.

The joining of structural elements in carpenter's fashion is superseded by the use of screws and bolts and later rivets. For the bridge over the river Brahe (today Brda) near Czersk, built in 1861 after a design by Schwedler, the truss joints were structurally designed as hinges (Figure 4a). Nine years later, another bridge over the Brahe was built near Bromberg (today Bydgoszcz), again designed by Schwedler, this time however using riveted joints (Figure 4b). In an 1865 essay structured into one hundred paragraphs, which was later repeatedly identified as a catechism of iron bridge construction, Schwedler stated that "the construction material of most iron bridges is rolled wrought iron" [Schwedler 1865: 333]. The disappearance of mixed systems from trussed framework construction not only signals the substitution of the carpenter by the metalworker, but also simplifies the theoretical treatment of iron construction practice through structural theory.

The iron trussed frameworks constructed in continental Europe after 1870 gradually became less reminiscent of the mechanical engineering tradition of early iron constructions: whilst in 1872 Emil Winkler (1835-1888) still provided detailed descriptions of both riveted and screwed truss joints, he recommended giving riveted joints preference over screw joints [Winkler 1872: 115]. Winkler was aware that the hinged trussed framework model contradicted the structural reality of the iron trussed framework with riveted joints; during the last decades of the 19th century, this contradiction led to the development of the theory of secondary stresses and – based on it – the method of deformations.

About the theory of secondary stresses

The truss members converging in the riveted joints are not only subjected to tension or compression through axial forces, but also to bending moments

(Figure 5); the latter generate bending stresses, which Friedrich Engesser (1848-1931) summarised under the term *secondary stresses* [Steinhardt 1949: 13]. The hinged trussed framework model only allows the stresses resulting from axial forces to be determined, and the quantification of the secondary stresses requires rather cumbersome calculations, since the structural system of the trussed frameworks with rigid joints is statically indeterminate. As early as 1872 Winkler stated that, according to his calculations, the secondary stresses can amount to up to 30% of the primary stresses resulting from the hinged trussed framework model [Winkler 1872: 115]. However, the first publications on the subject did not appear until 1879.

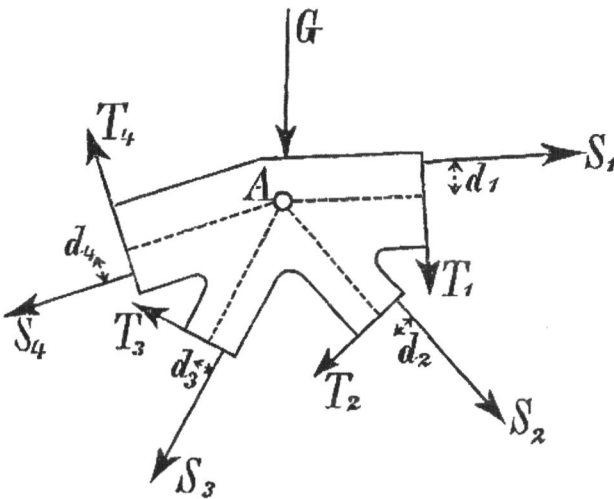

Figure 5. Internal forces at the rigid centric joint of a trussed framework [Winkler 1881a: 297]

In 1878, Heinrich Manderla (1853-1889), scientific assistant at Munich Technical University, submitted the complete solution for the competition set up by Professor Johann Gottfried Asimont (1834-1898): What stresses are generated in members of a trussed girder by the fact that the angle pieces of the trussed framework triangles are experiencing deformations caused by the load? Manderla's solution, which appeared in the 1878-79 annual report of Munich Technical University and was made available to a larger audience of experts as an essay in the *Allgemeine Bauzeitung* published in Vienna in 1880 [Manderla 1880], enabled the calculation of the secondary stresses in simple trussed frameworks with rigid joints, based on the second order theory [Kurrer 1985:

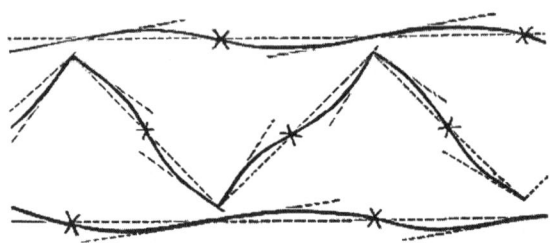

Figure 6. Trussed framework with eccentric joint [Winkler 1881a: 297]

a)

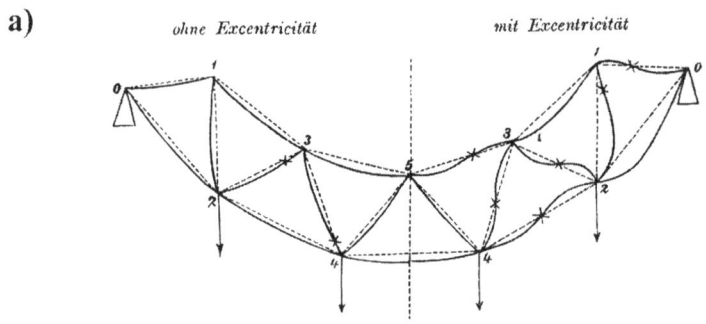

b)

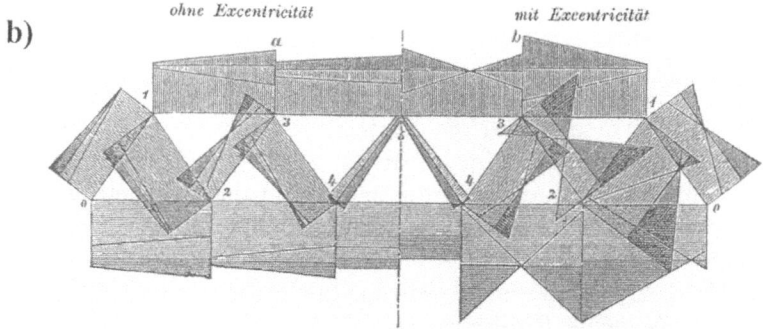

Figure 7. Analysis of the a) deformations and b) secondary stresses within a trussed framework with rigid joints without and with eccentricity [Winkler 1991a: 312]

327-328]. With Clebsch having first introduced unknown displacement parameters into mathematical elasticity theory for the calculation of truss systems in 1862, Manderla did the same in 1879 for structural theory.

In the same year, Engesser published an approximation method for determining the secondary stresses [Engesser 1879]. Engesser neglected the bending stiffness of the web members and analysed the top and bottom chords as continuous beams with imaginary supports in the joints. Engesser was well versed in the basic principles and may well have known Clebsch's book on elasticity theory, but there is no mention of Clebsch's method of deformations in any of Engesser's contributions on the theory of secondary stresses. The same applies to Emil Winkler, a true expert in the literature on basic principles, who made his name through comprehensive contributions on the theory of secondary stresses. Following Manderla's line of thinking, Winkler introduced the difference between end tangents and member chord angles of rotation in the joint, resulting in k linear equations for k joints from k moment equilibrium conditions [Winkler 1881b]. In contrast to Manderla, Winkler used the first order theory for the member moments. Winkler thus simplified the problem of secondary stresses to a fully linear problem, with the superposition principle applicable without restriction. On the other hand, he also considered eccentric truss joints (Figure 6), thus making the calculation more complicated again. In one of the many trussed frameworks analysed by Winkler, the increase in stresses compared with those calculated from the hinged trussed framework model was 14% on average for centric joints (Figure 7a) and 20% on average for eccentric joints (eccentricity e = 5 cm) (Figure 7b).

Apart from Manderla, Engesser and Winkler, Wilhelm Ritter (1847-1906), Theodor Landsberg (1847-1915) and Heinrich Müller-Breslau also made contributions to the theory of secondary stresses in the 1880s. After Georg Christoph Mehrtens (1843-1917), Schwedler too had his own method, although unfortunately he never published it. Using this method, during a six-week period in 1887 Schwedler calculated the secondary stresses in the main girders of the second bridge across the river Vistula at Dirschau and handed the results to Mehrtens for use in the design of this bridge [Mehrtens 1912: 226].

The keystone in the development the theory the secondary stresses was set down by Otto Mohr (1835-1918) with his work *Die Berechnung der Fachwerke mit starren Knotenverbindungen* (*The calculation of trussed frameworks with rigid joint connections*) [Mohr 1892-93]. Mohr's important contribution consisted in the clear differentiation of the joint angles of rotation ξ and the member angles of rotation ψ for the unambiguous determination of the deformation state of the trussed framework with rigid joints. Mohr developed his procedure based on a simple trussed framework with k = 7 joints (Figure 8). In the first step, using the

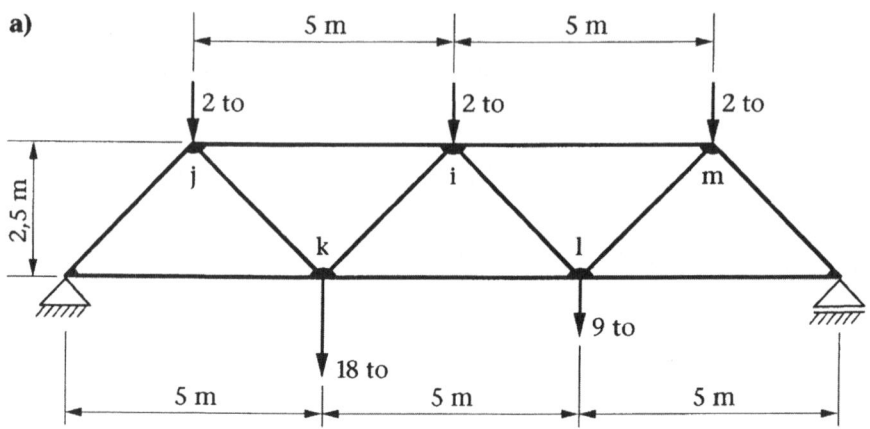

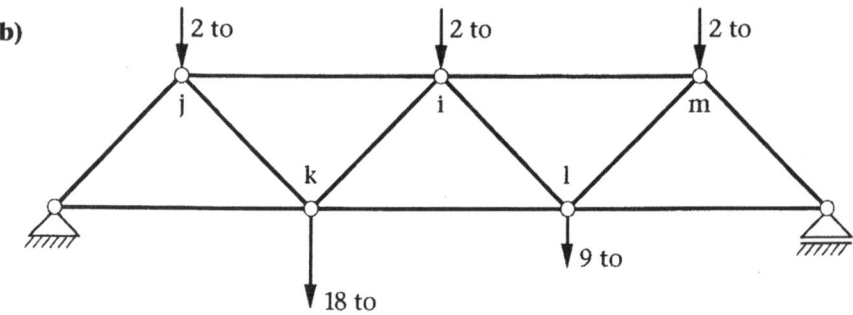

Figure 8. Example of an analysis after Mohr; a) trussed framework with rigid joints and b) associated hinged trussed framework model

principle of virtual forces, which in 1874-75 he introduced into the trussed framework theory (independent of James Clerk Maxwell (1831-1879)), Mohr calculated the member angles of rotation of the hinged trussed frameworks under load ψ_{ij}, ψ_{ik}, ψ_{il} and ψ_{im} (Figure 8b). Subsequently, the member end moments are determined for every joint, e.g., for joint i the moment:

$$M_{ij} = 2(EI_{ij}/l_{ij})(2\xi_i + \xi_j - 3\psi_{ij}) \qquad (3)$$

$$M_{ik} = 2(EI_{ik}/l_{ik})(2\xi_i + \xi_k - 3\psi_{ik}) \qquad (4)$$

$$M_{il} = 2(EI_{il}/l_{il})(2\xi_i + \xi_l - 3\psi_{il}) \qquad (5)$$

$$M_{im} = 2(EI_{im}/l_{im})(2\xi_i + \xi_m - 3\psi_{im}) \qquad (6)$$

From the moment equilibrium for joint i,

$$M_{ij} + M_{ik} + M_{il} + M_{im} = 0 \qquad (7)$$

an equation with the unknown joint angles of rotation ξ_i, ξ_j, ξ_k, ξ_l and ξ_m and the already-calculated member angles of rotation ψ_{ij}, ψ_{ik}, ψ_{il} und ψ_{im} can be derived.

For the specified trussed framework, Mohr noted the moment equilibrium conditions for every joint (7); he thus obtained a linear system of equations for determining the $k = 7$ joint angles of rotation. Having solved the linear system of equations, Mohr calculated the member end moments for every joint.

Mohr solved the problem of secondary stresses with a clarity and operative elegance that can hardly be surpassed, and at the same time created the foundation for the method of deformations. However, Mohr's contribution to the method of deformations disappeared from the focus of attention of structural theory for two decades for several reasons. First, the energetic imperative for structural theory advocated by Müller-Breslau was stronger than the kinematic imperative advocated by Mohr. Second, the force method resulting from the energetic imperative concluded the discipline formation period of structural theory and, for the time being, did not tolerate a second method in parallel. Third, Mohr published his article in a journal that was not very well known. Finally, Mohr was too focused on the trussed framework theory, which probably prevented him from generalising his work to include elastic truss systems.

From trussed frameworks to frameworks

In 1910, Willy Gehler (1876-1953) summarised the development of the theory of secondary stresses in his monograph *Die Entwicklung der Nebenspannungen eiserner Fachwerkbrücken und das praktische Rechenverfahren nach Mohr* (*The development of the secondary stresses in iron truss bridges and the practical calculation procedure after Mohr*) [Gehler 1910]. He reported comprehensively about measurements carried out at the railway bridge across the river Elster along the line Dresden-Elsterwerda just before Elsterwerda station. Gehler concluded that Mohr's procedure for determining the secondary stresses provided results "that show fully satisfactory consistency with the values observed in reality" [Gehler 1910: 69]. At the time Gehler made his observations, frameworks were already used in iron and reinforced concrete construction.

Since 1897, several iron bridges had been built in Belgium according to the Vierendeel system (see [Busse 1912] and [Vierendeel 1911]). The Vierendeel

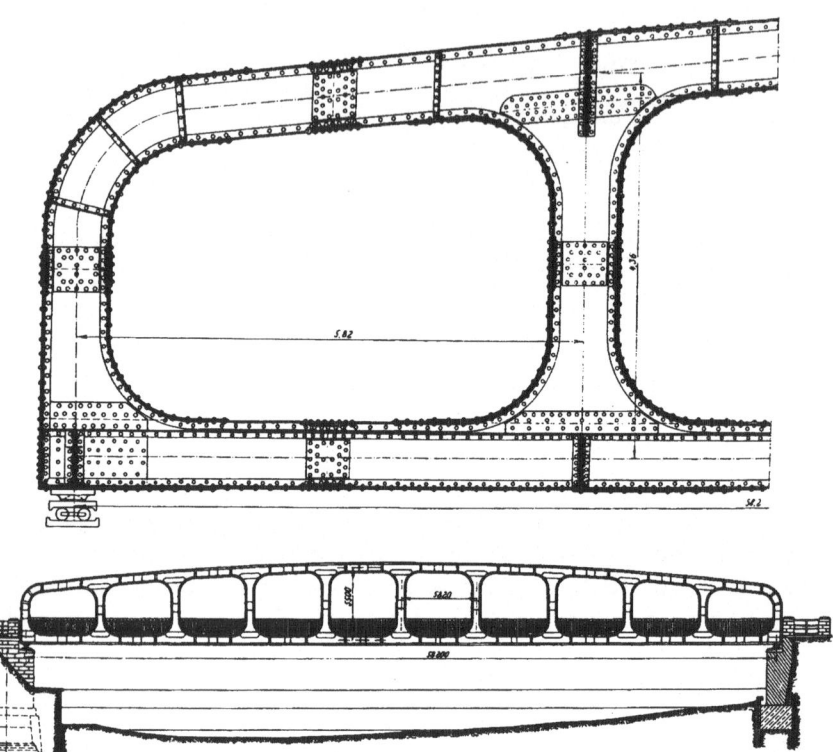

Figure 9. Design for a road bridge in the shape of a Vierendeel girder across the river
Ems near Salzbergen (never actually built) [Busse 1912: 215]

girder is a trussed framework without inclined bars, i.e., it is not made up of
triangles, but of rectangles (Figure 9). The Vierendeel girder was therefore also
called *post trussed framework*. In contrast to a member triangle, a hinged
member rectangle is kinematically simple: it has to be stabilised by additional
bars; the quadruple kinematic system shown in Figure 10b therefore requires
four additional bars to form a statically determined hinged system or stabilised
hinged system (Figure 10c).

Otto Mohr considered the reliability of the calculation of trussed frameworks
with rigid joints to be higher than that for Vierendeel girders. For this reason, he
advocated lower permissible stresses for the latter, which would make the
Vierendeel girder less economic than the trussed girder [Mohr 1912: 96]. In his
reply, Vierendeel rejected Mohr's objections and asserted that these applied to
trussed frameworks [Vierendeel 1912]. However, Mohr would be proved right

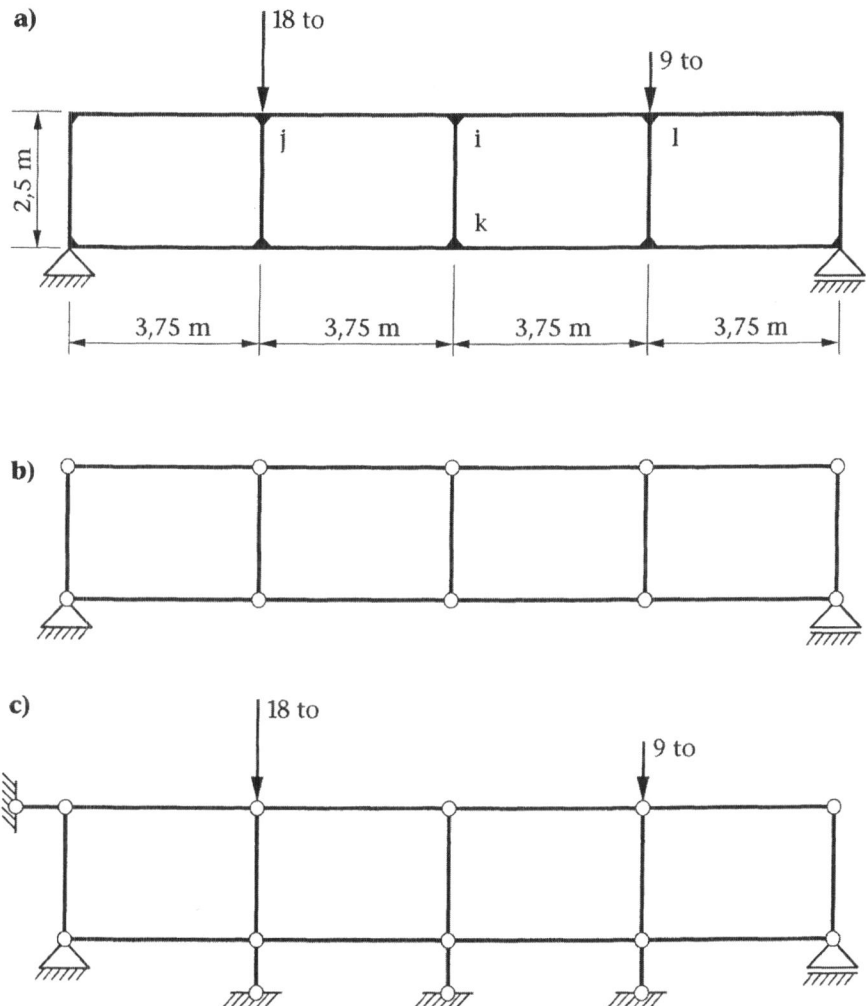

Figure 10. Vierendeel girder: a) static system; b) kinematic hinged system; c) stabilised hinged system

about the reliability of the calculation of Vierendeel girders. He was justified in criticising the simple beam model commonly used to calculate the internally statically-indeterminate Vierendeel girder. After 1910, numerous articles appeared that tried to redress these shortcomings and to master the internal static indeterminacy of the Vierendeel girder, amongst them Mohr's already mentioned work of 1912 [Mohr 1912]. The majority of authors used the force method, although this led to confusingly complicated algorithms that were unsuitable for practical structural calculations.

Nevertheless, in a historically-logically accentuated essay about the Vierendeel girder in iron construction published in 1912, Franz Czech developed numerous arguments for the Vierendeel girder and concluded that both theoreticians and practicioners could no longer ignore it [Czech 1912: 113]. According to the essay, the Vierendeel girder "had become established as the most economic structure" in reinforced concrete construction [Czech 1912: 113].

Rigid joints are an essential feature of reinforced concrete skeleton construction, which was invented by François Hennebique (1842-1921) and became established after 1900. Its monolithic character precluded modelling as a hinged system *a priori*. Whilst the trussed framework was the ideal load-bearing structure for classic iron construction, as early as 1910 the frame was seen as the appropriate load-bearing structure for reinforced concrete. The framework and frame analysis thus developed predominantly against the background of the penetration of civil engineering with the reinforced concrete construction between 1910 and 1920. During this time, a comprehensive body of literature about frame analysis was generated, for example the books by Willy Gehler [Gehler 1913] and Adolf Kleinlogel (1877-1958) [Kleinlogel 1914]: the focus of structural theory moved from the object world of iron construction to the object world of reinforced concrete construction.

The emancipation of the method of deformations from the trussed framework theory

The emancipation of the method of deformations from the trussed framework theory was the moment of the emancipation of structural theory from the object world of iron construction. Whilst the method of deformations did incorporate the theory of secondary stresses that stemmed from iron construction, up to the 1920s it mainly developed as structural theory of reinforced concrete.

In industrial construction, but also for the construction of long bridges, orthogonal frames were strung together horizontally (Figure 11). Vertical and horizontal sequences of orthogonal frames lead to multiple-span multi-storey frames, which in the 1920s and 1930s became the structural synonym for high-rise building construction (Figure 12). The force method kept reaching its limits for the analysis of such systems, because the linear system of equations grew with the number of statically-indeterminate parameters n. Structural theory therefore developed in two directions: the Berlin school of structural theory around Müller-Breslau [Kurrer 2000a] preferred the rationalisation of the force method, e.g., through orthogonalisation techniques [Kurrer 2000b], whilst technical

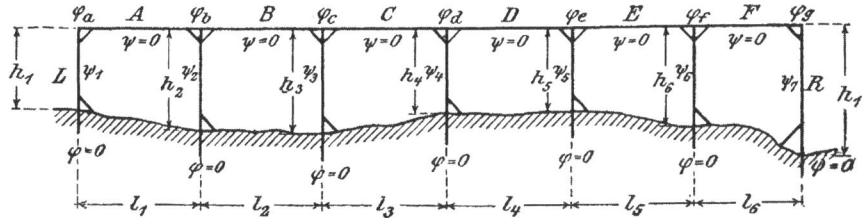

Figure 11.. Horizontal frame sequence [Gehler 1916: 103]

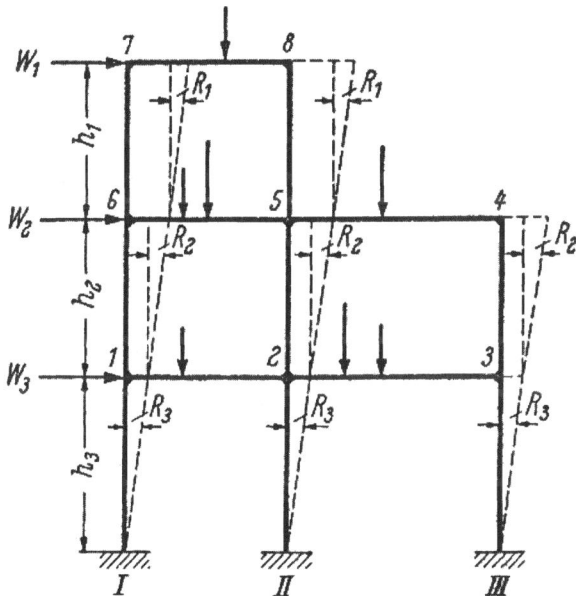

Figure 12. Vertical and horizontal frame sequence [Takabeya 1967: 46]

scientists, influenced by the Dresden school of technical mechanics around Mohr [Kurrer 2000b] were more inclined towards the method of deformations.

The reduction of the size of the linear system of equations for statically-indeterminate systems was the most important driving force for the further development of the method of deformations. Table 1 shows the number of equations according to the force method n and the method of deformations m for trussed girders with rigid joints (Figure 8a), Vierendeel girders (Figure 10a) and orthogonal frames (Figure 12) with p members and k joints. With the exception of the Vierendeel girder, m is invariably smaller than n. It is thus apparent that the method of deformations is particularly suitable for the structural analysis of frame sequences. For the Vierendeel girder, the force method is more advantageous, since the resulting system of equations will invariably contain two unknowns less than the system of equations for the method of deformations. For the case of trussed girders with rigid joints, Mohr had already demonstrated the superiority of the method of deformations, since the number of unknowns is reduced by 2k-6.

Structural system	n	m	n-m	n=m
Trussed framework (Fig. 8a: k=7, s=11	$3k - 6$ 15	k 7	$2k - 6$ 8	$2k = 6$
Vierendeel girder (Fig. 10a): k=10, s=13	$s - 1$	$s + 1$ 14	-2 -2	$n < m$
Orthogonal frames (Fig. 12): k=8, s=13	$3(s - k)$ 15	$3k - s$ 11	$4s - k$ 4	$2s = 3k$

Table 1. Comparison of the number of the unknowns of the forsce method (n) and deformation method (m)

Whilst the linear system of equations for trussed girders with rigid joints only contains joint angles of rotation ξ as unknowns, non-braced frames also contain member angles of rotation ψ or joint displacements w. Whilst trussed girders with rigid joints yield the stabilised hinged system almost automatically (Figure 8b), non-braced frames (Figure 10a) result in kinematic hinged systems (Figure 10b) that have to be stabilised by additional bars (Figure 10c). This forms the logical core of the emancipation of the method of deformations from the dependence of trussed framework theory.

Axel Bendixsen

In, 1914 the Danish engineer Axel Bendixsen extended Mohr's procedure [Mohr 1892-93] to braced and non-braced frames [Bendixsen 1914]. Bendixsen's approach had two stages (Figure 13):

The first stage yields α equations of loading from the analysis of the braced frame; these are the equations for the calculation of the joint angles of rotation α_i' for i = 1,...,6 (Figure 13b). Apart from the introduction of the support members, the first stage is identical to Mohr's procedure. In the supports, the forces Z_{10} and Z_{20} are generated (Figure 13b).

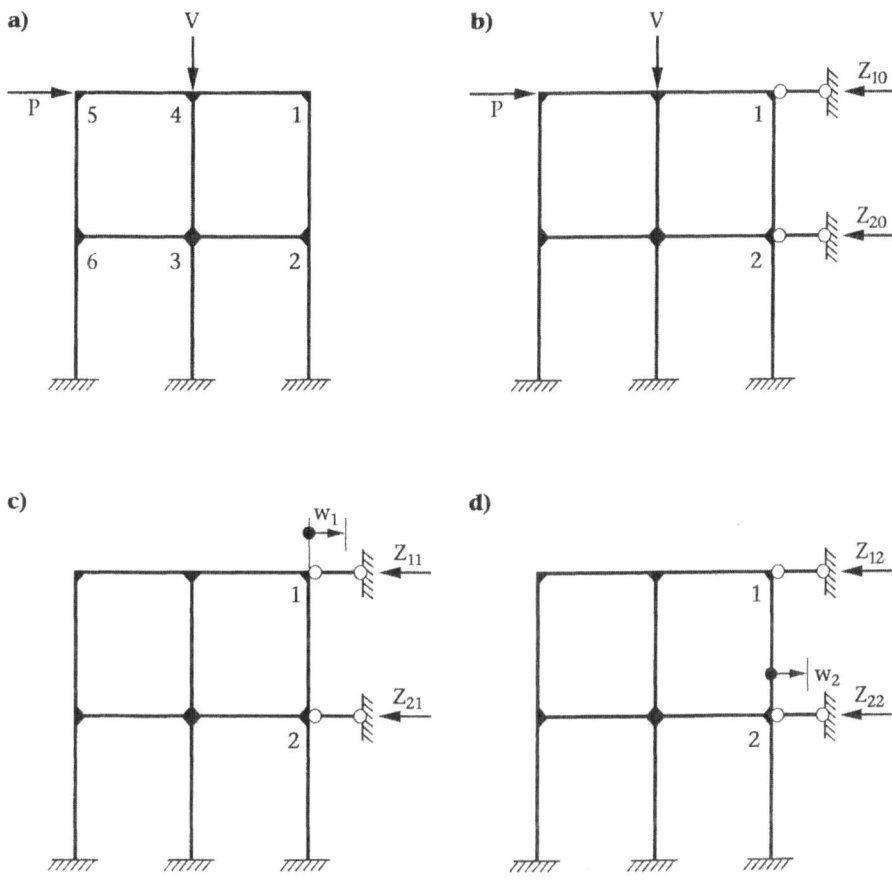

Figure 13. Multi-storey frame: a) in the loaded condition; b) with supports in the loaded condition in joints 1 and 2; c) with supports and impressed displacement w_1 or d) w_2

In the <u>second stage</u>, Bendixsen formulated the α *equations of displacement* for calculating the joint angles of rotation α_i'' for $i = 1,...,6$ by sequentially adding the unknown joint displacements w_1 and w_2:

$$\alpha_i'' = a_{i1}w_1 + a_{i2}w_2 \text{ for } i = 1,...,6 \qquad (8)$$

These equations can also be used to express the joint angles of rotation α_i'' due to the member angles of rotation. The second stage was completely new, despite the fact that it is formally similar to the force method and uses unit force states instead of unit displacement states. Joint displacement $w_1 = 1$ results in the reaction forces Z_{11} and Z_{21} (Figure 13c), joint displacement $w_2 = 1$ results in the reaction forces Z_{22} and Z_{12} (Figure 13d). The superposition of both unit displacement states leads to the reaction forces in

$$\text{joint 1: } Z_1'' = Z_{11}w_1 + Z_{12}w_2 \qquad (9)$$

and

$$\text{joint 2: } Z_2'' = Z_{21}w_1 + Z_{22}w_2 \qquad (10)$$

Bendixsen then determined the member angles of rotation due to w_1 and w_2 and inserted these values into the *α equations of displacement*. From the resulting joint angles of rotation α_i'' and member angles of rotation, the reaction forces Z_{11}, Z_{12}, Z_{22} and Z_{21} are calculated via the joint equilibrium conditions (here only $\Sigma H = 0$). The sums of the reaction forces from the *α equations of loading* and the *α equations of displacement* must be zero, since in reality joints 1 and 2 are not supported:

$$\text{joint 1: } Z_{10} + Z_1'' = 0 = Z_{10} + Z_{11}w_1 + Z_{12}w_2 \quad (11)$$

$$\text{joint 2: } Z_{20} + Z_2'' = 0 = Z_{20} + Z_{21}w_1 + Z_{22}w_2 \quad (12)$$

From these two equations, Bendixsen determined the joint displacements w_1 of joint 1 and w_2 of joint 2. Through evaluation of the appropriate *α equations of displacement*, the joint displacements yield the joint rotations α_i''. The final joint rotations ξ_i are calculated from the sum of α_i'' and the appropriate *α equations of loading* α_i':

$$\xi_i = \alpha_i' + \alpha_i'' \text{ for } i = 1,...,6 \qquad (13)$$

Only two significant technical journals responded with reviews and comments to Bendixen's pioneering contribution to the method of deformations. Bendixsen himself used his method of deformations for the calculation of ribbed dome structures [Bendixsen 1915].

Willy Gehler

Gehler, originating from the inner circle of the Dresden school of technical mechanics, published his angle of rotation method in 1916 [Gehler 1916]. He derived the member end moments M_{ij} as a function of the difference of joint and member angles of rotation: having satisfied the joint equilibrium conditions, he established a system of equations with m unknown joint angles of rotation ξ and member angles of rotation ψ. According to Gehler, one advantage of his method of deformations is that "the structural part of the problem is separated from the purely mathematical problem" [Gehler 1916: 104]. Finally, he points out the reduction in the number of unknowns and stresses the clear character of the unknown deformation parameters ξ and ψ, which can be measured directly at the structure.

Asger Ostenfeld

Around 1920, Asger Ostenfeld, professor at the Technical University of Copenhagen, extended Bendixsen's method of deformations in parallel with the force method. As early as 1921 he created the dual concept of the method of deformations [Ostenfeld 1921]:

- method of deformations versus force method;

- m unknown displacement parameters ξ_j versus n unknown force parameters X_i;

- reaction forces Z_{ij} versus difference of displacements δ_{ji};

- m linear equations for ξ_j versus n linear equations for X_i.

Ostenfeld noted the latter in the following form (following Einstein's summation convention):

$$-Z_{i0} = Z_{ij} \cdot \xi_j \text{ for } i,j,...,m \qquad (14)$$

$$-\delta_{j0} = \delta_{ji} \cdot X_i \text{ for } j,i,...,n \qquad (15)$$

In Ostenfeld's method of deformations, the inconsistencies inherent in Bendixsen's approach (namely, the application of Mohr's procedure [Mohr 1892-93], the separate determination the joint and member angles of rotation, and the two-stage quantification of the joint and member angles of rotation) were completely eliminated.

The significant progress in Ostenfeld's method of deformations lies in the fact "that it enables previously analysed structures, structural elements so to speak, to be built on" [Ostenfeld 1921: 288]. Ostenfeld achieves this not only, like Bendixsen, through the introduction of support members for obstructing joint displacements, but through the introduction of rigidly fixed members for the obstruction of joint rotations. This enables the complete frame to be subdivided into finite elements. Ostenfeld concludes:

> It can therefore be expected that this method will allow even rather complicated systems to be treated without difficulty, since, unlike with the force method, it is not necessary to start from scratch every time [Ostenfeld 1921: 288-289].

Using Ostenfeld's approach, Mohr's equation for the member end moment M_{ij}, for example, is simplified into the elementary cases of bars fixed at both ends (Figure 14):

$$\xi_i = 1: M_{ij} = 4(EI_{ij}/l_{ij}) \qquad (16)$$

$$\xi_j = 1: M_{ij} = 2(EI_{ij}/l_{ij}) \qquad (17)$$

$$\psi_{ij} = 1: M_{ij} = -6(EI_{ij}/l_{ij}) \qquad (18)$$

This leads to equation (16), if $\xi_i = 1$, $\xi_j = 0$ and $\psi_{ij} = 0$ is used in Mohr's equation (3) – the other two equations can be derived analogously. What may appear at first glance as an unnecessary complication will soon turn out to be an important contribution to the rationalisation of structural analysis.

In his book *Die Deformationsmethode* (*The deformation method*) [1926], Ostenfeld summarised his articles, which had appeared in the technical journals *Ingeniören (Engineering)* (1920, 1922), *Der Eisenbau (The iron construction)* (1921) and *Der Bauingenieur (The civil engineer)* (1923).

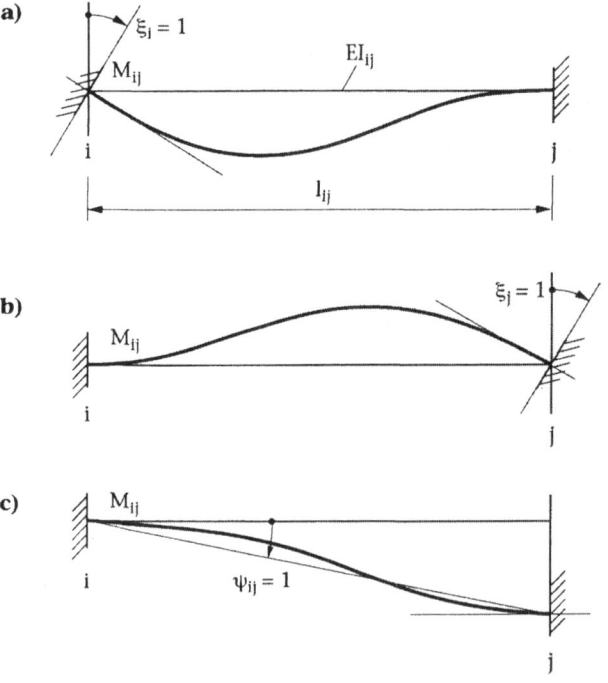

Figure 14. Elementary displacement load cases of the fixed bar: a) $\xi_i = 1$; b) $\xi_j = 1$; c) $\psi_{ij} = 1$

Ludwig Mann

One year after Ostenfeld's monograph, Ludwig Mann (1871-1959) published his book *Theorie der Rahmenwerke auf neuer Grundlage* (*A new basis for the theory of frameworks*) [Mann 1927]. Mann elaborated the method of deformations against the background of Joseph Louis Lagrange's *Mécanique analytique*. For the unknown displacement parameters ξ, Mann introduced the notion of base co-ordinates, a specification of Lagrange's generalised coordinates, with the first group of base co-ordinates representing the joint angles of rotation, the second group representing the member chord angles of rotation and the third group representing the independent elongations of the member chords. For calculating the reaction forces Z_{ij}, Mann used the principle of virtual displacements for the first time. Here too he followed Lagrange, who had based his complete mechanics on this principle. Finally, Mann called the system of equations for calculating the displacement parameters *Type 2 elasticity equations* and that for calculating the statically-indeterminate parameters *Type 1 elasticity equations*. In 1939, he extended his method of deformations to the calculation spatial truss systems [Mann 1939]. Mann became confronted with spatial truss systems in the context of calculations for large open-cast brown coal mining equipment in eastern Germany. For his work on brown coal mining, which was of crucial importance to the energy sector of the German Democratic Republic (GDR), founded in 1949, he was awarded the title of Dr.-Ing. E. h. (honorary doctor) by the Technical University Dresden in 1957. In a speech held on the occasion of the award ceremony, Mann developed the duality of force and deformation methods, based on the principle of virtual forces and the principle of virtual displacements, in classic clarity [Mann 1958].

With Mann's logical development of the force method in parallel to the method of deformations and his foundation of rational mechanics, the emancipation of the method of deformations from the trussed framework theory was completed. Historically, this completion constitutes the mid-point of the consolidation period of structural theory (1900-1950), which followed the formation period (1820-1890).

Method of deformations and modern structural mechanics

From 1930 to 1950, the method of deformations developed in three directions. Firstly, force and deformation method were compared with each other, and also different representations of the method of deformations. In 1934, Kruck examined the methods of Bendixsen, Ostenfeld and Mann together with Bleich's *Viermomentensatz* (*four moment law*) [Kruck 1934]. In 1937, Kruck developed Mann's difficult-to-comprehend method of deformations further to the *method of base co-ordinates* and formalised it through tables in such a way that the member parameters could be inserted directly into the summation terms of the basic equations [Schrader 1969: 8].

Secondly, further insight was gained into the dual nature of structural theory; relevant contributions were made by Pasternak in 1920 [Pasternak 1920] and 1926 [Pasternak 1926] and by Hertwig in 1933 [Hertwig 1933]. With the clear abstract distinction of the principle of virtual forces from the principle of virtual displacements, based on the variation principles of the elasticity theory [Schleusner 1938], the method of deformations caught up with the force method in terms of its formal structure.

Thirdly, the method of deformations opened up areas of application that significantly exceeded those of structural theory. Examples are truss dynamics (see [Fliegel 1938] and [Koloušek 1941]) and the theory of stability, which was heavily influenced by metal construction practice. It was particularly the analysis of components with risky stability from the aeroplane, machines and steel construction industries that required the influence of deformations on the equilibrium to be taken account of. The formulation of such a second order theory in the language of the force method would be very cumbersome, whereas the method of deformations poses no such problems, as shown by Chwalla and Jokisch in 1941 [Chwalla and Jokisch 1941], and later by Teichmann [Teichmann 1958].

The emancipation of the method of deformations from structural theory, and the laying of its foundations in the shape of the two complementary principles of mechanics, historically form the content of the preparation phase of the development of modern structural mechanics after 1950. The implementation of matrix calculus in structural aviation engineering, in structural theory and in technical mechanics after 1950 finally led to the synthesis of those disciplines into modern structural mechanics.

Whilst the force method, translated by Levy, Lang, Bisplinghoff, Langefors, Wehle and Lansing into the language of matrix calculus between 1947 and 1952, found its way into the analysis of complicated aeroplane structures [Kurrer 2000b: 197], in 1956 it was criticised by Turner, Clough, Martin and Topp:

> The method is, of course perfectly general. However, the computational difficulties become severe if the structure is highly redundant, and the method is not particularly well adapted to the use of high-speed computing machines [Turner, Clough, et al. 1956: 806].

In contrast, on the following page the creators of the finite element method paid tribute to the matrix formulation of the method of deformations, published by Levy in 1953 (stiffness method) [Levy 1953]:

In a recent paper, Levy has presented a method of analysis for highly redundant structures which is particularly suited to the use of high-speed digital computing machines [Turner, Clough, et al. 1956: 807].

In 1957, Argyris translated the complete structural theory into the language of matrix algebra and, at this level, elaborated the duality of the force and deformation methods [Argyris 1957]. Argyris's *Matrizentheorie der Statik* (*Matrix theory of structural design*) was – like the foundation of the finite element method through Turner, Clough, Martin and Topp – a cornerstone of modern structural mechanics. The force method formulated in matrix notation, lost ground *vis-a-vis* the method of deformations, since the latter was far better suited for implementation on the computer. Indeed, in the 1960s there was no shortage of contributions that suggested using the method of deformations as the basis for a problem-oriented programming language [Schrader 1969].

The method of deformations became the centre of modern structural mechanics, and in the late 1960s became the preferred theory of the finite element method. Ostenfeld's hope, stated in 1926, that the elementarisation of trusses into finite elements would succeed in treating complicated systems without problem, thus became reality not only for one-dimensional, but also for multi-dimensional continua.

References

ARGYRIS, JOHN. 1991. Die Matrizentheorie der Statik. *Ingenieur-Archiv*, vol. 25, n. 3: 174-192.

BENDIXSEN, AXEL. 1914. *Die Methode der Alpha-Gleichungen zur Berechnung von Rahmenkonstruktionen*. Berlin: Springer.

———. 1915. Die Berechnung von Rippenkuppeln mit oberem und unterem Ringe. *Armierter Beton*, vol. 8: 45-49, 76-80, 95-101 and 114-119.

BENVENUTO, EDOARDO. 1991. *An Introduction to the History of Structural Mechanics. Part II: Vaulted Structures and Elastic Systems*. Berlin: Springer.

BUSSE, R. 1912. Entwurf einer Rahmenbrücke über die Ems. *Der Eisenbau*, vol. 3, n. 6: 214-219.

CHWALLA, ERNST and FRIEDRICH JOKISCH. 1941. Über das ebene Knickproblem des Stockwerkrahmens. *Der Stahlbau*, vol. 14, n. 8/9: 33-37 and n. 10/11: 47-51.

CLEBSCH, ALFRED. 1862. *Theorie der Elasticität fester Körper*. Leipzig: B.G. Teubner.

CULMANN, KARL. 1851. Der Bau der hölzernen Brücken in den Vereinigten Staaten von Nordamerika. *Allgemeine Bauzeitung*, vol. 16: 69-129.

CZECH, FRANZ. 1912. Der Vierendeelträger in der Geschichte des Eisenbaues. *Der Eisenbau*, vol. 3, n. 3: 104-113.

ENGESSER, FRIEDRICH. 1879. Über die Durchbiegung von Fachwerkträgern und die hierbei auftretenden zusätzlichen Spannungen. *Zeitschrift für Baukunde*, vol. 2: 590-602.

FLIEGEL, E. 1938. Die Elastizitätsgleichungen zweiter Art der Stabwerksdynamik. *Ingenieur-Archiv*, vol. 9, n. 1: 20-38.

GEHLER, WILLY. 1910. *Die Entwicklung der Nebenspannungen eiserner Fachwerkbrücken und das praktische Rechnungsverfahren nach Mohr.* Berlin: Wilhelm Ernst & Sohn.

————. 1913. *Der Rahmen. Einfaches Verfahren zur Berechnung von Rahmen aus Eisen und Eisenbeton mit ausgeführten Beispielen.* Berlin: Wilhelm Ernst & Sohn.

————. 1916. Rahmenberechnung mittels der Drehwinkel. in: Willy Gehler (ed.), *Otto Mohr zum achtzigsten Geburtstage.* Berlin: Wilhelm Ernst & Sohn: 88-123.

HERTWIG, AUGUST. 1933. Das „Kraftgrößenverfahren" und das „Formänderungsgrößenverfahren" für die Berechnung statisch unbestimmter Gebilde. *Der Stahlbau*, vol. 6, n. 19: 145-149.

KLEINLOGEL, ADOLF. 1914. *Rahmenformeln.* Berlin: Wilhelm Ernst & Sohn.

KOLOUŠEK, VLADIMIR. 1941. Anwendung des Gesetzes der virtuellen Verschiebungen und des Reziprozitätssatzes in der Stabwerksdynamik. *Ingenieur-Archiv*, vol. 12: 363-370.

KRUCK, GUSTAV E. 1934. Beitrag zur Berechnung statisch unbestimmter, biegungsfester Tragwerke. Zürich: PhD-Thesis, ETH Zürich.

KURRER, KARL-EUGEN. 1985. Zur Geschichte der Theorie der Nebenspannungen in Fachwerken. *Bautechnik* vol. 62, n. 10: 325-330.

————. 2000a. Die Berliner Schule der Baustatik. Pp. 152-163 in: *1799-1999. Von der Bauakademie zur Technischen Universität Berlin.* Karl Schwarz, ed. Berlin: Ernst & Sohn.

————. 2000b. Von der Kunst zur Automation des statischen Rechnens. Pp. 188-199 in: *1799-1999. Von der Bauakademie zur Technischen Universität Berlin.* Karl Schwarz, ed. Berlin: Ernst & Sohn.

LEIPHOLZ, HORST. 1968. *Einführung in die Elastizitätstheorie.* Karlsruhe: G. Braun.

LEVY, S. 1953. Structural analysis and influence coefficients for delta wings. *Journal of the Aeronautical Sciences*, vol. 20: 449-454.

LORENZ, WERNER. 1990. Die Entwicklung des Dreigelenksystems im 19. Jahrhundert. *Stahlbau*, vol. 59, n. 1: 1-10.

LORENZ, WERNER. 1997. 200 Jahre eisernes Berlin. *Stahlbau*, vol. 66, n. 6: 291-310.

LOVE, AUGUST EDWARD HOUGH. 1907. *Lehrbuch der Elastizität*. German edition translated by Aloys Timpe. Leipzig/Berlin: B. G. Teubner.

MANDERLA, HEINRICH. 1880. Die Berechnung der Sekundärspannungen, welche im einfachen Fachwerk in Folge starrer Knotenverbindungen auftreten. *Allgemeine Bauzeitung*, vol. 45: 27-43.

MANN, LUDWIG. 1927. *Theorie der Rahmenwerke auf neuer Grundlage*. Berlin: Springer.

———. 1939. Grundlagen zu einer Theorie räumlicher Rahmentragwerke. *Der Stahlbau*, vol. 12, n. 19/20: 145-149 and n. 21/22: 153-158.

———. 1958. Vergleich der Prinzipien und Begriffe für die Entwicklung der Kraft- und Deformationsmethoden in der Statik. *Bauplanung - Bautechnik*, vol. 12, n. 1: 12-15.

MEHRTENS, GEORG CHRISTOPH. 1912. *Vorlesungen über Ingenieur-Wissenschaften I. Teil: Statik und Festigkeitslehre, Dritter Band II. Hälfte*. 2nd edition. Leipzig: Wilhelm Engelmann.

MOHR, OTTO. 1892/93. Die Berechnung der Fachwerke mit starren Knotenverbindungen. *Zivilingenieur*, vol. 38: 577-594 and vol. 39: 67-78.

———. 1912. Die Berechnung der Pfostenträger (Vierendeelträger). *Der Eisenbau*, vol. 3, n. 3: 85-96.

OSTENFELD, ASGER. 1921. Berechnung statisch unbestimmter Systeme mittels der „Deformationsmethode". *Der Eisenbau*, vol. 12, n. 11: 275-289.

———. 1926. *Die Deformationsmethode*. Berlin: Verlag von Julius Springer.

PASTERNAK, PETER. 1920. Beiträge zur Berechnung vielfach statisch unbestimmter Stabsysteme. *Der Eisenbau*, vol. 13, n. 11: 239-254.

———. 1926. Der abgekürzte Gauss'sche Algorithmus als eine einheitliche Grundlage in der Baustatik. Zürich: PhD-Thesis, ETH Zürich.

RAMM, WIELAND. 2001. Über die Geschichte des Eisenbaus und das Entstehen des Konstruktiven Ingenieurbaus. *Stahlbau*, vol. 70, n. 9: 628-641.

SCHLEUSNER, ARNO. 1938. Das Prinzip der virtuellen Verrückungen und die Variationsprinzipien der Elastizitätstheorie. *Der Stahlbau*, vol. 11, n. 24: 185-192.

SCHRADER, KARL-HEINRICH. 1969. *Die Deformationsmethode als Grundlage einer problemorientierten Sprache*. Mannheim: Bibliographisches Institut.

SCHWEDLER, JOHANN WILHELM. 1851. Theorie der Brückenbalkensysteme. *Zeitschrift für Bauwesen*, vol. 1: 114-123, 162-173 and 265-278.

———. 1865. Resultate über die Konstruktion der eisernen Brücken. *Zeitschrift für Bauwesen*, vol. 15: 331-340.

STEINHARDT, OTTO. 1949. *Friedrich Engesser*. Karlsruhe: C. F. Müller.

TAKABEYA, F. 1967. *Mehrstöckige Rahmen*. Berlin: Wilhelm Ernst & Sohn.

TEICHMANN, ALFRED. 1958. *Statik der Baukonstruktionen III. Statisch unbestimmte Systeme*. Berlin: Walter de Gruyter.

TURNER, M. J., R.W. CLOUGH, H.C. MARTIN and L.J. TOPP. 1956. Stiffness and deflection analysis of complex structures. *Journal of the Aeronautical Sciences*, vol. 23: 805-824.

VIERENDEEL, A. 1911. Der Vierendeelträger im Brückenbau. *Der Eisenbau*, vol. 2, n. 10: 381-385.

————. 1912. Einige Betrachtungen über das Wesen des Vierendeelträgers. *Der Eisenbau*, vol. 3, n. 6: 242-244.

WINKLER, EMIL. 1872. *Die Gitterträger und Lager gerader Träger eiserner Brücken*. Wien: Carl Gerold's Sohn.

————. 1881a. *Theorie der Brücken. Theorie der gegliederten Balkenträger.* 2nd edition. Wien: Carl Gerold's Sohn.

————. 1881b. Die Sekundär-Spannungen in Eisenkonstruktionen. *Deutsche Bauzeitung*, vol. 14: 110-111, 129-130 and 135-136.

A timbrel dome in construction. The wooden planks at the top are not centering but serve only as a guide to control the geometry during the construction. Note the transverse walls on the springing; they may be used to build the roof, but they are also necessary from a structural point of view. East Boston High School, Boston, Massachusetts, 1899; Architects Brown and Moses; dome design and construction by Rafael Guastavino (Photo courtesy Avery Library, Columbia University)

THE MECHANICS OF TIMBREL VAULTS:
A HISTORICAL OUTLINE

Santiago Huerta[1]

Timbrel vaults are masonry vaults, with a good strength in compression, and can be constructed with remarkable thinness and without the use of formwork. Known in the fourteenth century and commonly constructed by the sixteenth, until the middle of the nineteenth century, timbrel vaulting was used vaulting in churches, floor systems and staircases. At the end of the nineteenth century Rafael Guastavino exported the method to the United States, where it was used in many important buildings. This paper examines the development of the theory of timbrel vaults from Espie in the eighteenth century, through Bails and Fornés in the nineteenth, to Guastavino and Guastavino, Jr. in the twentieth, to the use of Finite Element Methods (FEM) today.

Introduction

Timbrel vaults are masonry vaults made with brick and mortar. Their uniqueness derives from their construction: the bricks are placed flatly, forming one or more layers and they are constructed without centering or other support. The bricks are placed in arches or successive rings to complete the vault (Figure 1). During construction, the bricks are supported by the adhesion of the fast-setting mortar to the completed courses, or to the bordering walls. There is no formwork, but guides are used to control the geometry of the vault, particularly for large vaults or for a high-quality finish [Moya 1957; Gulli 2001].The method is analogous to the construction of brick vaults without formwork, which were widely used in Byzantium [Choisy 1883]. These are built with lime mortar, which sets very slowly, and the adhesion of the bricks is achieved by inclining the brick courses, but the construction proceeds in a similar way forming arches or rings. The coincidences suggest a common origin, but the question is still open to further research [González 1999; Mochi 2001; Tarragó 2001].

Timbrel vaults can be constructed with remarkable thinness. Normally two layers of brick tiles are used (about 10 cm in total thickness, including the mortar between the layers), but vaults of only one layer of brick can be found (5 cm thickness). The slenderness ratio, relating the radius of curvature to the span, is typically around 100, but many vaults are even thinner. Also, timbrel vaults with large spans have been built, the greatest being the dome over the crossing in St. John the Divine, New York, with 33 m [Ramazotti 2001].

---

[1] Departamento de Estructuras, ETS de Arquitectura, Universidad Politécnica de Madrid, Avda. Juan de Herrera 4, 28040 Madrid, SPAIN

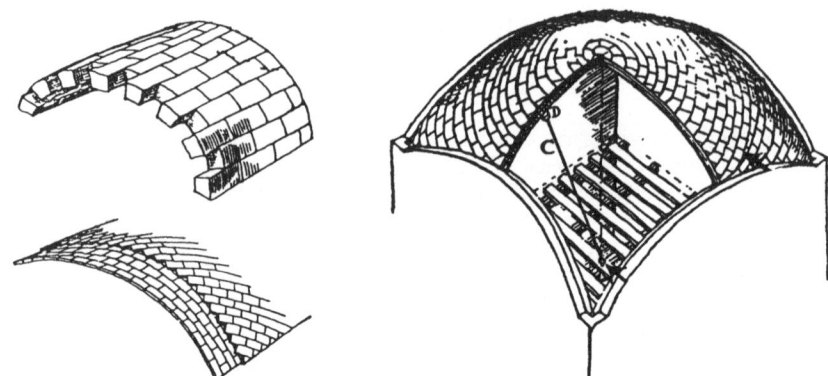

Figure 1. Construction of timbrel vaults. (a, left) Comparison between a timbrel vault
and a stone voussoir vault; (b, right) Construction without centering of a timbrel dome.
The geometry is controlled by a rod attached to a fixed point [Moya 1957]

Until the middle of the nineteenth century, timbrel vaulting was used for
various construction elements: a) to cover the naves of churches. In this case,
they could only support their own weight and the occasional loads due to
maintenance, and, in general, a timber roof protected these vaults; b) for floor
systems; c) for staircases. From the beginning of the nineteenth century, timbrel
vaulting began to be used in Spain and France for the construction of roofs and
floors of industrial buildings, principally textile factories. The use of Portland
cement allowed them to be used as roofs, without the need for an additional roof
or other waterproofing methods. In Catalonia towards the end of the nineteenth
century and the beginning of the twentieth century the timbrel vault became
something of a national symbol [Neumann 1999]. Rafael Guastavino, a Spanish
architect, exported the method to the United States at the end of the nineteenth
century, and the method developed a greater importance than ever before.
'Guastavino vaulting' was used in many of the most important buildings between
1890 and 1900 in the eastern United States [Collins 1968].

The method of timbrel vault construction is fairly well known and
comprehensive bibliographies can be found in the works of Collins [1968], Gulli
and Mochi [1995], González [1999] and Huerta et al. [2001]. But the same
cannot be said of the structural behaviour of timbrel vaults. The first
architectural treatises made no essential distinction between the structural
behaviour of timbrel vaults and that of brick or stone vaults. But, at the
beginning of the eighteenth century they were viewed by some architects with
scepticism, due to a perceived lack of durability and safety. In particular, timbrel
vaults were considered to function in a completely different manner than
conventional stone or brick vaulting; as we shall see, they were considered to be

monolithic and to exert no thrust. Guastavino classified them as "cohesive constructions," as opposed to structures held in place by gravity. Then followed attempts to make elastic analysis, which ended many times in failure. Eventually, in Spain, timbrel vaulting came to be known as "impossible to calculate," and as a result, some of them have been demolished and substituted with more conventional structural systems.

The primary objective of this article is to trace the history of the ideas concerning the structural behaviour of timbrel vaults and, finally, to return timbrel vaults to their place: timbrel vaults are masonry vaults.[2] Like any other masonry structure, they have little resistance to tension, they crack, and they thrust. They are neither monolithic nor cohesive. They can and should be calculated with the same methods used for a vault of masonry. They are also durable if they receive the necessary maintenance.

Traditional timbrel vault design in Spain: Fray Lorenzo de San Nicolás.

The first documents on this type of construction can be traced back to the fourteenth century [Araguas, 1999], and timbrel vaults of the same period still survive in Catalonia. In the sixteenth century they were commonly constructed [Marías, 1991]. They were valued for ease of construction, high strength, and above all, their lighter weight, which allowed for considerable reductions in the size of the supporting walls and buttresses. Explicit reference to the cited advantages can be found in some reports written ca. 1620 during the construction of the Palace of Carlos V in Granada [Rosenthal, 1988].

The most relevant text on the construction and mechanics of timbrel vaulting is the architectural treatise of Fray Lorenzo de San Nicolás, published in Madrid in 1639. Fray Lorenzo, who worked as an architect and built many timbrel vaults, describes the construction of the fundamental types of vaulting (barrel, groined, hemispherical, cloister, etc.) in stone, in brick with radial joints, and in timbrel vaulting. No distinctions are made as to whether one material is better or worse than another. Fray Lorenzo, it seems, considered the three methods to be equally good constructively and he left the selection to the architect in each case. Furthermore, it is revealing that, independent of the material, the vault must be provided with lateral support to carry the thrust to the buttresses. He indicates that it is necessary to fill the haunches for the first third of the vault height and to provide supporting transversal walls, called *lengüetas*, in the second third. (This is the traditional way of construction, which was followed more than two hundred years later by Guastavino; see the introductory figure.)

---

[2] The present article is a revised and modified version of Huerta [2001c].

Fray Lorenzo is explicit in the structural role of these elements:

> ...*and as you continue constructing, you will layer and solidify the haunches until the first third, and in all of the vaults, placing the* lengüetas, *which rise for another third, and in this manner they will receive the thrust or the weight of the vault* [Fray Lorenzo 1639: fol. 91v].

The fill and transverse walls or *lengüetas* could serve to support the horizontal thrust of soil, but they also had a structural function: they permitted the vault to resist asymmetrical overloads and moving loads.

Fray Lorenzo carried out the calculation of buttress sizes. He gave a series of rules that referred to the standard construction of the period: a church of one nave, sometimes with lateral chapels, covered with a barrel vault (the plan has the form of a Latin cross and over the crossing a dome is built, Figure 2). He proceeded in a systematic form, assigning the dimensions to the vault and considering two possible types of buttress: a continuous wall or a wall with counter-forts. His exposition is so systematic that can be summarized in Table 1.

TYPE OF VAULT	TYPE OF BUTTRESS		
	Wall (uniform section)	Wall with counter-forts	
		Wall thickness	Wall plus counterfort
Stone vault	1/3	1/6	≥1/3
Brick vault: radial joints	1/4	1/7	1/3
Brick: timbrel vault	1/5	1/8	1/4

Table 1. Fray Lorenzo's rules for buttress design. In the treatise the exposition is in the running text, but it is so systematic as to be presented in the form of a table

The ancient builders identified the thrust of the vault with the necessary buttress to resist it. The timbrel vault thrusts less than the brick vault or stone vault, but it thrusts, and requires a system to counter the thrust (Figure 2).

The treatise of Fray Lorenzo gained widespread diffusion in later centuries in Spain (it was still being used by builders at the start of the twentieth century). This is not surprising because the book is exceptional for the number of themes treated and for the clarity of its presentation. Its rules for buttresses are mentioned in many later architectural treatises, for example García Berruguilla

Figure 2. Typical longitudinal section of a Spanish parish church of the seventeenth century. Note the thinness of the dome shell, which may correspond to a projected timbrel dome [Fray Lorenzo 1639]

[1747] and Plo y Camín [1767].

Of course, Fray Lorenzo and the rest of the educated builders of timbrel vault construction knew that once a timbrel vault was finished, the only difference in the structural behaviour compared to the conventional brick or stone vault was the decreased thrust due to the lighter weight. They continued supporting the vaults with buttresses, although fewer were required. Everything else was identical. In particular, the timbrel vaults could be cracked and the pathologies would be identical to those of brick or stone vaults.

The timbrel vault tradition in France:
The Comte D'Espie and the myth of 'monolithism'

In France there existed a tradition of timbrel vaulting due to Spanish influence in the region of Rousillon, which Bannister [1968] has studied exhaustively. Around 1700 this tradition passed to the French Languedoc region and in particular, the Duke of Belle Isle built a series of timbrel vaults in his castle, employing the bricklayers of Perpignan. The construction of these lightweight vaults caused a great sensation at the time, and was discussed in the Académie Royale d'Architecture following a paper presented 19 June 1747 by M. Tavenot

[Lemmonier, 1920]. The Académie did not approve of this construction method, which was new to them, but the paper presented by Tavenot is relatively extensive and includes additional information in the appendices.

This type of construction excited the curiosity of a learned nobleman, retired by this point, the Comte d'Espie. He was interested in the possibility of building floors and roof systems in timbrel vaulting, due to its excellent fire resistance. He studied buildings that contained timbrel vaulting and finally constructed a building made with this fireproof construction system. He collected his experiences and opinions in a small book, published in 1754, titled "*Manière de rendre toutes sortes d'édifices incombustibles, ou Traité sur la construction des voûtes, faites avec des briques et du plâtre, dites voûtes plates, et d'un toit de brique, sans charpente, appelé comble briqueté*" (*Manner of constructing all sorts of fireproof buildings, or treatise on the construction of vaults, made with brick and plaster, called flat vaults, and of a roof of brick, without wood, named comble briqueté.*[3]) The book was well received and within a few years was translated and published in English [1756], German [1760], and Spanish [1776]. A second French edition followed in 1776.

Espie begins his book discussing the advantages of timbrel vaulting, stressing its fireproof quality, as well as its lightness and adaptability. He gives also a detailed description of the construction method. But what is of interest for the present study is that he dedicates a chapter to comparing timbrel vaulting with ordinary vaulting: "Parallel des Voutes ordinaires Avec des Voutes Plates" [Espie 1754: 40-58]. He starts by describing the qualitative form in which masonry vaults thrust against the supports. He discusses the influence of the thickness of the vault, its height or rise, and the height of the buttress. He cites Bélidor [1729] in relation to the calculation of thrusts and he warns of the danger of basing projects in practice alone, without some base in theory. Immediately he observes that the cited rules cannot be applied to timbrel vaulting because they are of a different nature and they do not thrust against the walls:

> *Les voutes plates étant d'une nature différente n'ont pas besoin qo'on suivre dans leurs constructions les mêmes regles & les mêmes principes que dans les précédentes; il est donc inutile d'examiner si les murs sont épais ou non. . . . car je ne suis pas de ceux qui croient que ces Voutes poussent les murs* [Espie 1754: 44].

Espie attributes this absence of thrust to the monolithic character of the finished vault, which forms a solid mass due to the good quality of the mortar (plaster) employed. Because it is impossible to form cracks and divide itself the

---

[3] Lemma [1996] includes a facsimile reproduction and an Italian translation.

vault exerts no thrust:

> *...car le Plâtre lorsqu'il est bien lié avec la Brique fait de toute la Voute entiére un corp massif qui n'a aucun jeu dans ses parties: elles ne se pousseront jamais les unes contre les autres, puisque le tout ensemble ne fait qu'une masse solide qui se contiendra toûjours d'elle-même sans se diviser, pour peu qu'elle soit soutenue* [Espie 1754: 57].

Then he gives a series of observations, made by him personally and others, in support of his monolithic no-thrust theory. In one case he made a load test; in another, he cut away the vault except for the four corners. He made holes in completed vaults to test their resistance. He also writes of a man who built a small vault over a wooden frame, and once the mortar was set, he rolled it around the ranch and hit it with a hammer. They are clearly the reflections of an 'amateur' outside of the tradition of timbrel vaulting. It is interesting to note the 'scientific' approach, trying to obtain conclusions from experiments, but the experiments can be interpreted in several ways and, in fact, many of the tests can be made with normal masonry vaults with the same results. No theory is a direct consequence of a series of experiments.

However, the ideas and experiences gathered by Espie were generally accepted without criticism by later authors. The absence of thrust and the resistance to fire were powerful assertions that created an immediate interest, not only in France but also in the rest of Europe. It was unusual for a work to be translated into Spanish, English and German so quickly. Furthermore, important French and European writers echoed the ideas of this 'new' construction system, basing their comments on Espie's book and relating his ideas. This is the case of Laugier (1755) and of Rieger [1763], but it was particularly important that timbrel vaulting received an extensive treatment in the writing of Blondel and Patte [1771-1777], one of the most influential works of the period. In the sixth volume they dedicated a complete chapter with 40 pages and 7 excellent plates. Undoubtedly, this greatly contributed to the spread of timbrel vaulting (Figure 3).

Two decades later, Rondelet [1802] summarized this information in a section of his *Traité de l'art de bâtir*, including drawings. The treatise of Rondelet was one of the most influential of the nineteenth century. It was translated into German and Italian, and numerous editions were printed. Therefore, at the beginning of the nineteenth century in France there was a theory of timbrel vault construction that was based primarily on the opinions of a learned nobleman who aimed "to serve the community."

In summary, some of the most outrageous ideas about timbrel vaulting (its monolithic nature, absence of thrust, etc.), which spread rapidly throughout Europe, had their origin in the treatise of the Comte d'Espie. These ideas formed the official 'frame of reference' for approaching these structures, and came back to Spain, then under a heavy French influence.

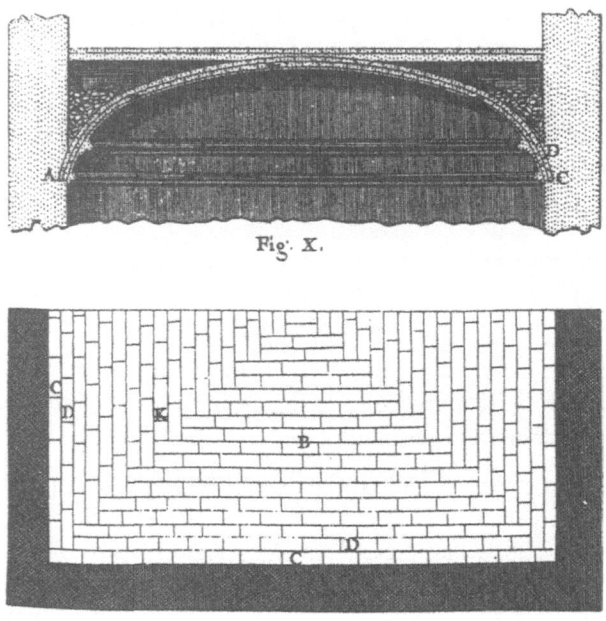

Figure 3. Construction a timbrel vault for the floor of a building. Note the filling on the haunches and the existence of transverse walls to support the floor. In plan the form of bonding of the bricks forming successive arches parallel to the walls [Blondel 1777: vol. 6, Plate 95]

The Spanish edition of Sotomayor and the 'Censure' of Ventura Rodríguez

Joaquín de Sotomayor [1776] translated the Espie's book into Spanish, adding his own opinions and experiences in brackets. More interestingly, the book was preceded by a 'Censure' from Ventura Rodríguez, the official architect of Madrid and one of the most important Spanish architects of the eighteenth century. The beginning captures the tone of the Censure: "We would obtain considerable advantages in the art of building if all of the feasible ideas we propose were as successful in practice as they seem in fantasy."

There follows a sharp critique of the fundamental ideas of Espie: the monolithic nature and the resulting lack of thrusts. Ventura Rodríguez cites various cases of cracking and displacements in completed buildings which demonstrate the thrust of the vaults:

> *Though this supposition [the absence of thrust], or belief, is flattering, it is in spite of the cited experience and cannot be verified, as affirmed by the evident examples that we have in almost all of the Temples of Madrid, whose vaults are timbrel vaults of brick and plaster, of high curvature, and with thick walls, protected by buttresses, whose firmness is a great advantage . . . and we have seen them broken in many places, and with the walls displaced, due to the thrust.*

He insists numerous times in the necessity of providing sufficient support for timbrel vaults and he emphasizes the importance of "firmness", in addition to the 'beauty' and the 'convenience', for if this fails, "all is lost". Clearly, Ventura Rodríguez does not agree with the opinions of Espie and considers them to be dangerous. In fact, Sotomayor, like Espie, was an 'amateur' of the construction method, but not a builder. Ventura Rodríguez, an architect of great experience, saw immediately the mistakes and perils of Espie's 'monolithic', no-thrust theory.

The first scientific experiments in France

Apparently the interest in timbrel vault construction continued in France during the nineteenth century. Historical studies on this theme are lacking and the only evidence we have found is the realization of experiments trying to ascertain the thrust and strength of timbrel vaults: D'Olivier [1837] and Fontaine [1865]. It is remarkable that in both cases the thrust of the vaults is considered (negating implicitly Espie's theory) and that the calculations were made following conventional masonry vault theory. Of particular interest are the large-scale tests to failure described by Fontaine. One of the tests described is on three timbrel vaults with a span of 4 m (and a rise of 0.4 m), spanning between wrought iron I-beams (of 47 cm depth) with a span of 6.25 m, covering a total area of 72 sq m. The test was carried out until failure occurred under a load of 1,250 kg/m^2. In another test on a timbrel vault spanning 3.75 m (again with rise:span ratio of 1:10), the vault carried a load of 2,700 kg/m^2 without failing. Tests of such magnitude were not made in an isolated manner and, most probably, were carried out in the hope of producing fireproof vaults for factories. (In fact the size of the test of the three vaults coincides with the usual plan module for textile factories.)

Spanish treatises of the first half of the nineteenth century: Bails and Fornés

In nineteenth-century Spain, Espie's influence is evident in two later Spanish treatises which treat timbrel vaulting, those of Benito Bails [1796] and Manuel Fornés [1841, 1846]. Bails basically compiled and plagiarized previous French manuals, particularly Blondel/Patte. He dedicates a chapter to timbrel vaulting. Initially he transcribes the corresponding paragraphs of Fray Lorenzo de San Nicolás, but then he copies, translating into Spanish, directly from Blondel/Patte. He is apparently unaware of the contradiction between both texts.

The writing of Fornés is original. First published in 1841 and revised in 1857, Fornés presents the method of building timbrel vaults and makes a new contribution. He sets out in great detail the way of building the principal types of timbrel vaulting: barrel vaulting, staircases, domes, squinch arches, etc. In regard to the thrust, Fornés considers that timbrel vaults thrust, though less, due to their smaller thickness, following the traditional ideas of Fray Lorenzo and Ventura Rodríguez.

However, Fornés knows the ideas of Espie, probably through Bails, and contradictions begin to appear. Thus, in the first part of his treatise he affirms he discusses the geometry and thickness of the vaults from which originates the thrusts and the size of the walls to resist it. But further on he writes: "[the timbrel vault]…covering the work and walls, the material becomes a solid body, equal to the lid of a pot, with no more thrust than its weight" [Fornés 1841: 47]. As with Bails, Fornés seems unaware of his contradictions. However all the projects of vaulted buildings included in his second treatise possess the usual buttressing of masonry buildings.

Rafael Guastavino's theory of 'cohesive construction'

Rafael Guastavino was the first to attempt to formulate a theory that explained, in a scientific form, the structural behaviour of timbrel vaults. To put his work in context it is necessary first to give a brief biographical sketch of his fascinating life. Born in Valencia in 1842, he went to Barcelona in 1861 where he began his studies of building and architecture. By 1866 he had already built apartment house and in 1868 began the construction of the huge Batlló Factory. There he used the timbrel vault technique extensively (Figure 4), and by then he was convinced that the future progress and perfection of masonry construction would be with this type of building. This idea became the objective of his life, his *Leitmotiv*.

Afterwards he built a number of buildings in Barcelona [Bassegoda 2001]. Then, he won an Award in the Philadelphia Exhibition of 1876 and eventually

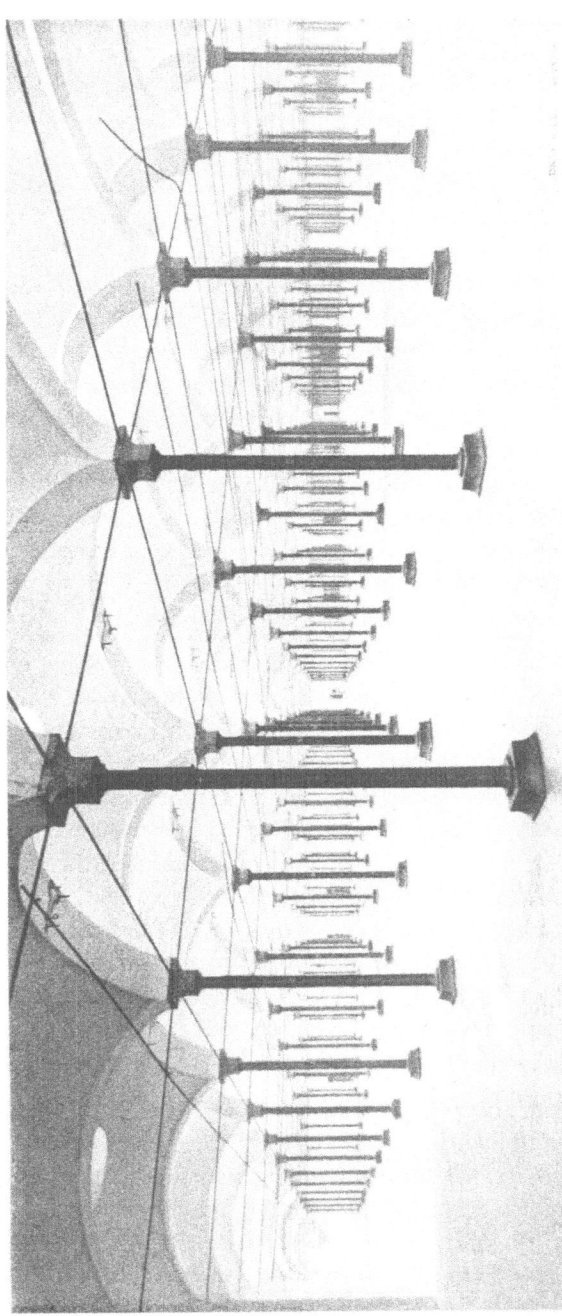

Figure 4. Batlló Factory in Barcelona (1868-1870). View of the interior with an extensive use of timbrel domes (Archivo Histórico de la Diputación de Barcelona)

decided to immigrate to America, arriving at New York in 1881 with his elder son Rafael Guastavino, Jr. After a brief stint as an architect, he decided that the best way to promote the use of timbrel vaulting was to work as a building contractor and in 1889 he established the Guastavino Fireproof Construction Company. The same year he began his first great contract: the building of the vaults of the Boston Public Library (McKim, Mead and White, architects). The lightness and audacity of this new structure aroused great admiration. It should be kept in mind that, before Guastavino, many of the vaults in American buildings were 'false', hanging from a wooden or iron structure; this was cheaper than the usual stone or brick vaults. Timbrel vaults, lighter and constructed without the need of heavy centering, were attractive to many architects. After the Boston Public Library, Guastavino worked for some of the most important architects of this time [Collins 1968]. But it was not at all easy; this type of vault was completely unknown in America and was looked with suspicion for many builders. The first task of Guastavino was, then, to convince American architects and engineers of the strength and high quality of these structures, and also to demonstrate their kinship with great masterworks of architecture, such as the Pantheon in Rome or Hagia Sophia.

Guastavino needed a theory both technical and historical, and he had been conscious of this since his first works in Barcelona. He first presented his ideas at a series of seminars for the Society of Arts at the Massachusetts Institute of Technology in 1889. He also published a series of magazine articles in 1889, and, finally, presented the ideas in a book, *Essay on the Theory and History of Cohesive Construction, applied especially to the timbrel arch*, published in 1892 and reprinted with minor corrections in 1893. (For the genesis of this book see Parks [2001]). Guastavino later published additional articles and conference papers (for a complete bibliography see [Huerta et al. 2001]), as well as another book titled *Prolegomenos on the use of masonry in modern constructions* [1896-1904]. While this last book is fundamental to understanding the architectonic thinking of Guastavino, it does not include new information on his ideas and calculations pertaining to the structural behaviour of timbrel vaulting. Therefore in what follows we will refer to the *Essay* of 1893.

The first part of the *Essay* is autobiographical and in it Guastavino explains the sources of his thoughts. He mentions the classes he received in the School of Architecture in Barcelona from his instructors, Juan Torras and Elías Rogent. Guastavino acknowledges them for calling his attention to this method of construction, which, he later affirms at various places in his book, had been forgotten for a long period of time. This last assertion is extremely doubtful.[4]

---

[4] This assertion is more than debatable. Timbrel vaulting had had a constant presence in the Spanish construction treatises since the seventeenth century. It is also a significant fact that an

We do not know the content of these classes in Barcelona, but it is likely that the professors presented Espie's ideas of continuity and monolithic nature, which were known in the Spanish treatises of the time [Sotomayor 1776; Bails 1796; Fornés 1841].

Guastavino divided masonry structures into two groups according to their structural behaviour:

We will divide construction in general into two classes:

First, "Mechanical Construction," or construction by gravity.

Second, "Cohesive Construction," or construction by assimilation.

The first, is founded in the resistance of any solid to the action of gravity when opposed by another solid. From these conjunctive forces, more or less opposed to one another, results the equilibrium of the total mass, without taking into consideration the cohesive power of the material set between the solids.

The second has for a basis the properties of cohesion and assimilation of several materials; which, by a transformation more or less rapid, resemble Nature's work in making conglomerates [Guastavino 1893: 45].

Timbrel vault construction is cohesive, but it is not the only type of cohesive construction. In the second half of his *Essay* he makes a confusing historical revision. Roman concrete construction is 'evidently' cohesive, but Guastavino also considered Byzantine and Islamic brick constructions as cohesive; besides, he claims the Middle Ages as the era in which the cohesive system truly developed. Also the great domes of the Renaissance are cohesive. In fact, the list of cited buildings includes some of the most notable buildings from different periods and styles: the baths of Caracalla, the Hagia Sofia in Istanbul, the Cathedral of Zamora, the great Cathedral and Baptistery of Florence, St. Peter's in Rome, Sainte-Geneviève in Paris, St. Paul's in London, and two timbrel vaulted domes in Valencia [Guastavino 1893: 26-29]. Apparently, any building constructed in a material with good adhesion of the mortar, including Roman concrete, brick, timbrel vaulting, etc., falls in the category of cohesive construction. No doubt

---

important part of the treatise by Fornés, published in Valencia in 1841 (2nd ed. 1857), is dedicated to timbrel vaulting. Fornés systematically uses timbrel vaulting in the construction specifications of his "Album de proyectos" of 1846. It seems then, that timbrel vault construction was well known in Valencia in the middle of the nineteenth century, not to mention the extraordinary timbrel domes which existed in the city. As for the rest of Spain, the treatise of Ger y Lóbez, published in Badajoz in 1869, also discussed timbrel vaulting, to the same extent that it discussed vaulting in brick or stone.

Guastavino is looking for historical arguments in favour of timbrel vault construction.

Another point of great importance is his remark about the 'natural' character of cohesive timbrel vault construction. Guastavino was fascinated by the possibility of constructing a solid from many small pieces, in the same way that nature forms conglomerates. He describes his fascination with a visit to the great cave of the Monasterio de Piedra in Spain:

> *The thought entered my mind, while in this immense room . . .*
> *that all this colossal space was covered by a single piece, forming a*
> *solid mass of walls, foundation and roof, and was constructed with*
> *no centres or scaffolding . . .*
>
> *This grotto is really a colossal specimen of cohesive construction.*
> *Why had we not built on this system?* [Guastavino 1893: 13;
> (emphasis mine)]

This passage is the key to understanding the structural thinking of Guastavino. The idea that cohesive construction (including timbrel vaulting) is a 'natural' construction, and furthermore, was 'more rational, durable, and economical', came to him as a revelation and was a driving force for his work throughout his life. As we will see, the cohesive character does not influence the essential behaviour of timbrel vault structures, but the research to improve the cohesion led to an unprecedented perfection of timbrel vault construction.[5] On the other hand, the observation about the monolithic character of structure ("covered by a single piece") echoes the ideas of Espie, which he may have heard in his classes in Barcelona. However, the essential character of timbrel vault construction, the possibility of dispensing with centering, though mentioned, became a matter of secondary importance in the *Essay*.

Advantages of the 'cohesive' timbrel vault

Guastavino explained the differences between gravity construction and cohesive construction in relation to timbrel vaults, aiming to prove the advantages of the latter [Guastavino 1893: 49-57]. He compares a timbrel arch of one layer with a timbrel arch composed of two layers (Figure 5). In the single-

---

[5] This is not unusual in the history of the development of science and technology. From time to time, a scientist or artist is led by a false idea, but their enthusiasm helps them to develop other correct ideas which can lead to an advance in the discipline. Koestler (1964) cites, among other examples, the case of Kepler, who had a lifelong obsession with the geometrical harmony of the movement of spheres, discovering laws which shattered the Greek geometrical ideas on the movement of stars.

Figure 5. Comparison between a 'mechanical' arch (above) and a 'cohesive arch' (below) [Guastavino 1893]

Figure 6. Construction of timbrel arches. The man standing is Rafael Guastavino, demonstrating the strength of these thin arches. Boston Public Library, 1889-1890, McKim, Mead and White, architects (Avery Library, Columbia University)

layer arch there are joints between the bricks, which, he says, function like the voussoir in a traditional gravity arch. The double-layered arch, with mortar between the two layers and with overlapping joints, forms an arch which functions as a cohesive structure, capable of resisting bending moments. The evidence for this assertion is that it is possible to construction barrel vaults spanning twenty ft (6 m) with a thickness of only 3 in (7.5 cm). After only a few hours, the workers can walk on top of the vault safely – and this is indeed a proof of a certain bending resistance. Finally, the form used in construction can be placed as proof that the vault has not deformed. Guastavino attributed great importance to this characteristic of timbrel arches, and it is no coincidence that he photographed himself standing on one of the recently completed timbrel arches of the Boston Public Library (Figure 6).

Guastavino attributed many of the structural advantages of timbrel vaults and arches to the reduction in the number of joints. If it were possible to construct without joints it would be ideal: "It is evident that if we were able to build an arch without joints, it would be the best, as it would have no settlement" [Guastavino 1893: 52]. Once again, Guastavino cites the myth of a monolithic nature. Of course masonry arches and vaults crack due to changes in geometry; this is the only way in which the masonry structures adjust to changes in the boundary conditions.[6]

Guastavino summarizes the advantages of the timbrel arches and vaults in relation to mechanical arches:

- The vertical joints are protected from cracking by the overlapping of joints;

- There are fewer vertical joints;

- There is capacity to resist bending moments.

Of course, massive concrete arches (without reinforcing) are cohesive arches with no vertical joints, but Guastavino discards them due to excessive cost and problems with irregular setting of the concrete [Guastavino 1893: 56].

Load tests

Guastavino was very aware of the problem of convincing the American architects of the merits of timbrel vault construction. Even in Spain, where the

---

[6] The phenomenon is well known since antiquity. The first interpretations of the cracking as a result of support movements appeared in the middle of the nineteenth century. Jacques Heyman made the first systematic studies; see, for example, Heyman [1997].

method had been used for centuries, these structures were often viewed with a lack of confidence. Guastavino's theoretical and historical speculations were necessary, but above all, he had to make scientific tests. Although earlier tests were made in France (see above), Guastavino was unaware of them and made his own tests. The first systematic tests were made in 1887 on timbrel specimens (Figure 7). Later, in 1901, he carried out also structural load tests, and fire tests to demonstrate the strength and "fireproof" nature of timbrel vaulting.

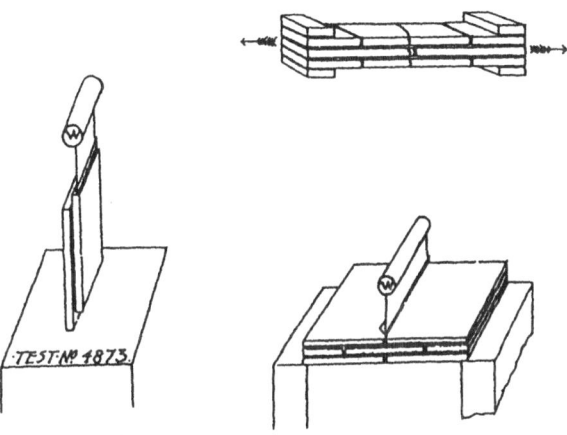

Figure 7. Specimens for the strength tests made by Guastavino: (a, upper right) tension; (b, lower right) bending ; (c, left) shear. [Guastavino 1893]

In the material tests he tried to obtain breaking stress values for compression, tension, shear, and bending. These values could then be used to verify the safety of his vaults by comparing the working stresses with the material failure stress. This of course, was the focus on stress and strength which began with Navier [Heyman 2001]. The results of the tests are summarized in Table 2. It is interesting that he does not mention any attempt to determine the elastic constants, such as Young's modulus or Poisson's ratio.

STRENGTH	N/mm^2
Compression	14.60
Tension	2.00
shear	0.90

Table 2. Mean strength of timbrel specimens [Guastavino 1893]

Figure 8. Load test made by Guastavino on a barrel vault, 1901 (Avery Library, Columbia University)

Figure 9. Load tests made by Guastavino in a staircase under construction, 1903
(Avery Library, Columbia University)

To compare these results with the real structures, he made failure tests of flat timbrel vaults with a rise of 1/10 of the span. One of these tests is shown in Figure 8. The photo is spectacular, and says more in favour of the strength of timbrel vaults than any theory or laboratory test results. In this case, it is clear that these tests played a role as propaganda, which was common at the end of the nineteenth century.[7]   On other occasions, load tests were made during construction (Figure 9).

The thrust of timbrel vaults and domes

Practically the only information on the calculation methods used by Guastavino is found in his *Essay*. (The graphic methods of structural analysis can be attributed to Guastavino, Jr., and will be discussed later.) Guastavino treated two typical themes: the flat barrel vault and the hemispherical dome, also flat. To obtain the thrust of a flat arch or barrel vault he gives the following formula (I have modernized the notation) [Guastavino 1893: 59]:

---

[7] It is interesting to compare the tests and the pictures with those made by Hennebique on reinforced concrete. See Delhumeau [1999].

$$A(S_{br}) = \frac{WI}{8f} \qquad [1]$$

where A = cross-sectional area of the vault at the crown per unit length; S_{br} = breaking stress in compression; W = total load (self-weight plus fill and live load) acting on the vault per unit of length; I = span of the vault; f = rise of the vault.

The formula relates the load W with the area A (depth) and the breaking stress of the material for an arch of given geometry. Guastavino treats the equation for the thrust of a parabolic arch under a uniformly distributed load as a given, though his 'demonstration' is difficult to understand. The formula is approximate since the loading is not exactly uniform, but for flat vaults it is sufficiently accurate. Guastavino gives an example application: to calculate the thickness of a vault spanning 15 ft (4.575m), with a rise/span ratio of 1/10, under a uniform load of 250 lbs/ft^2 (12 kN/m^2). The material has a breaking strength of 2120 lbs/in^2 (14.6 N/mm^2). Guastavino considered the permissible working stress to be 1/10 the breaking stress, so the allowable working stress would be 212 psi (1.46 N/mm^2). Entering these values in the formula gives a required thickness of 1.85 in (4.7 cm), which requires two layers of one-inch thick bricks. Of course, a safety factor of ten is excessive even for an irregular material like the masonry of timbrel vaulting. Further on, Guastavino admits that working stresses can be considered at 1/4 or 1/5 the breaking stress [Guastavino 1893: 64].[8] In fact, in masonry structures and in timbrel vault structures, the criterion governing safety is not the strength of the material but the stability of the system. Safety is obtained by giving a sufficient thickness. Perhaps the oscillation between four and ten for the coefficient of safety allowed Guastavino to choose the thickness that seemed adequate to him in each case.

The formula gives the thickness at the crown. The force is larger at the supports and to find the new thickness would require the application of "Dejardin's formula" [Dejardin 1860], stating that the force increases from the

---

[8] The factor of safety of 10 applied to masonry has an origin. The strength of a masonry element depends on the strength of the stones, the form and dimensions of the mortar joints, and the strength of the mortar, as shown by the tests of Tourtay [1885]. Thus, to derive factors of safety against breaking, one must consider blocks of a certain size. However, the first tests were made with small pieces of stone. Knowing that the strength of the actual element would be much less, the engineers of the nineteenth century took, in an empirical form, the admissible strength of the fabric as 1/10 the breaking strength of the stone. Of course, the rule does not apply to the tests made on small samples, as in the case of Guastavino. This fact led to considerable confusion in the engineering manuals at the end of the nineteenth century, and many times, to absurdly low values of admissible stress in masonry structures.

crown down to the support.[9]   Guastavino's calculation is evidently an equilibrium calculation. It obtains a value of thrust, and later checks this value against the strength of the material (normally unnecessary), and also to calculate the value of lateral thrust at the supports. The thrust was usually resisted with masonry buttresses, or more frequently, using a system of wrought iron ties.

However, at the end of the nineteenth century the elastic theory was considered as the best option to analyse masonry arches, and the equilibrium method, although used in practice, was viewed with suspicion by engineers. In all likelihood, Guastavino did not have sufficient training to make an elastic calculation, which required the solution of complicated integrals for even the simplest cases. For this reason he hired a professor of applied mechanics from MIT, Gaetano Lanza [1891], to calculate a table of the elastic stresses in timbrel arches, taking into account the normal force and the bending moment. The table is included at the end of the book without any explanation. This was another attempt to give scientific respectability to the calculation of timbrel vaulting. Comparing the results of his formula with the elastic table, there are not significant differences, but this is logical for flat arches.

Guastavino moves on to domes, which he considered to be an excellent form: "The dome is the genuine form of cohesive construction for ceilings, floors, and roofs, as well as for timbrel arches" [Guastavino 1893: 66]. To calculate the thrust, Guastavino makes another approximation, reasoning (incorrectly) in a geometrical manner to compare the areas of a sphere and a half cylinder of the same radius developed in plan. In effect cutting the cylinder as indicated in the figure and joining the wedges forms a polygonal dome approximating a sphere. Seeing the plan, Guastavino considered that the weight of the dome is one half the weight of the corresponding barrel vault, and therefore, the thrust would be half. In fact, the weight is different and also changes the position of the centres of gravity, but to consider the thrust of a dome as half the thrust of a vault is a safe assumption, since the thrust is normally smaller (for a hemispherical dome the thrust is close to one third). The idea comes from Frézier [1760: 3, 406], later spread through some architectural manuals, and Guastavino could have learned

---

[9] Dejardin's book was very popular in the second half of the nineteenth century, and it included rules and observations of practical interest. The rule to obtain the thickness variation of an arch is very old. It originates in the equilibrium analysis by La Hire [1695] of a semicircular arched formed by rigid voussoirs without sliding. To maintain equilibrium, the condition is that the weight of the voussoirs (i.e., thickness) must vary with the inverse of the cosine. Frézier was the first to propose arches of variable section on the basis of this. Later it became common in masonry arch bridges, but not in arches for building construction [Huerta, 1990]. In the case of a timbrel vault, which supports moderate loads, it does not seem necessary to give such low working stresses. However, the choice is logical if one considers that it is a problem of material strength (though this is not the case) as Guastavino did.

this in his classes in Barcelona. For flat arches, Guastavino's formula [1] gave a good approximation of the thrust, but in this case there could be significant variations, as Guastavino himself recognized: "We do not pretend to give an absolute mathematical formula, but a practical one, which is sufficient to guarantee the building safety" [Guastavino 1893: 68, 72].

Guastavino is using the usual, simple formulae for the thrust of masonry, voussoir, arches, vaults and domes, but he is apparently unaware of this. On the other hand, after calculating the thrust in the manner indicated, without considering bending moments, he contradicts himself: "We consider our arch not as an arch with voussoirs, but as a single cast arch, working as a solid piece of arched stone or iron" [Guastavino 1893: 69]. He later states, "We are here considering the dome as not one of voussoirs, but as a simple cast dome working as a single piece." [Guastavino 1893: 72]. He then dedicates several paragraphs to explaining that a timbrel arch thrusts, but less than a voussoir arch, and continuing the reasoning applied to domes he observes that cohesive domes, which resist tension, are made of their own rings and as a consequence the dome never thrusts. However, in any constructive section of a Guastavino dome there is a metallic hoop to resist this thrust. Numerous drawings of such metallic elements exist in the Guastavino archives and the tension rings are present in every dome section of the Guastavino files in the Avery Library.

The treatise contains many other observations on the structural functioning of timbrel vaulting, and some of them demonstrate a deep understanding of the structural behaviour of timbrel vaults. But the text is also, as we have seen, filled with inconsistencies as a result of wanting to apply, in any case, his cohesive theory. After giving the reader an enlightening observation about an aspect of the structural behaviour of some element, he begins again with his contradictions and his dubious assertions.

Theory and practice in the work of Guastavino

Guastavino's cohesive theory, often incorrect and with numerous contradictions, leads one to ask: how is it possible that he was one of the greatest builders of masonry vaults and domes? The enormous variety of constructive solutions, the genius and mastery exhibited, the audacity to design unprecedented domes, are all in stark contrast to the primitive character of his theory. How is this possible?

On one hand, Guastavino began from the reference point of monolithic behaviour, cohesion, tensile and bending strength: all marks of the dominant mode of thinking in the time period in which he lived. The second half of the nineteenth century was the period of the development of elastic theory, and it

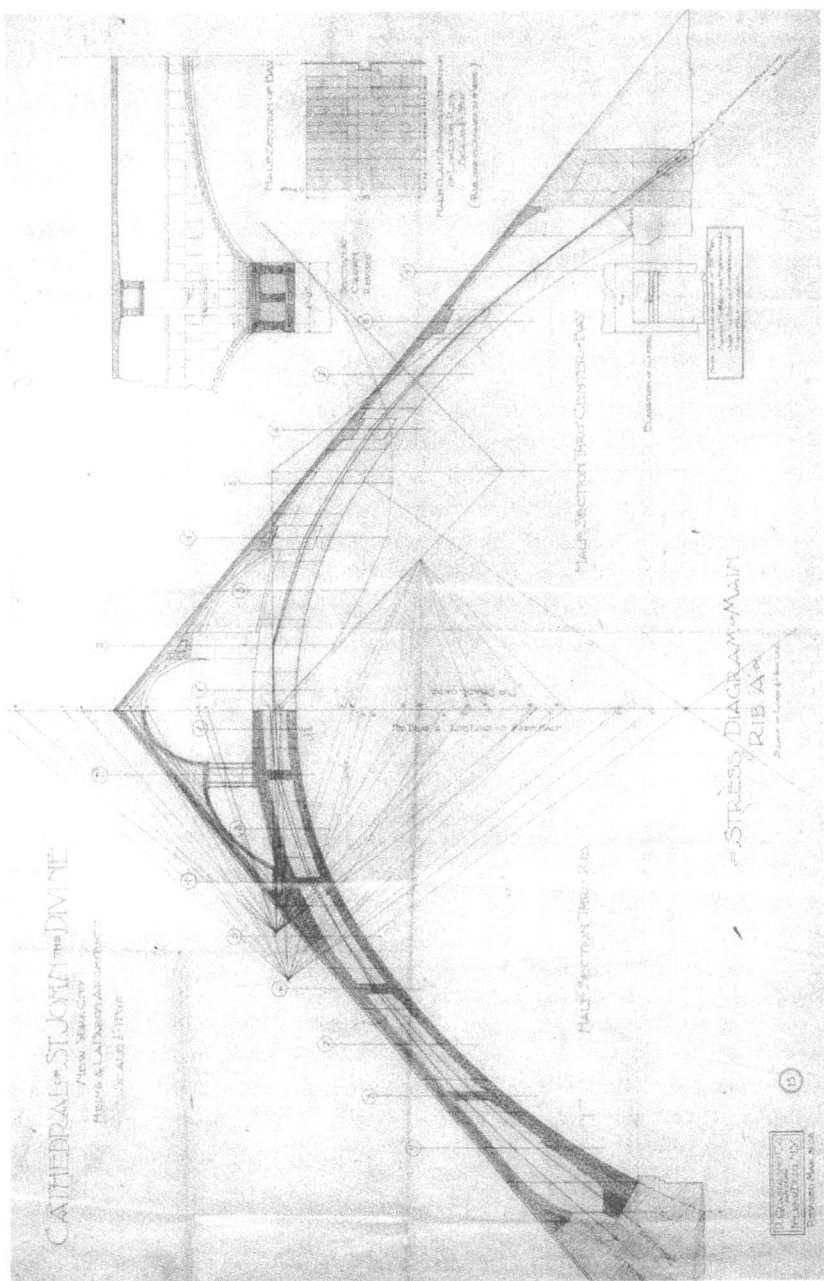

Figure 10. Graphical equilibrium analysis of the projected hollow timbrel arches for the roof of the nave of St. John the Divine (1892-1932). In the end, the arches were not constructed and the nave was covered with metallic trusses (Avery Library, University of Columbia)

easily incorporated the earlier concepts of Espie (changing monolithic behaviour for continuity, homogeneity, isotropic materials, etc.).

On the other hand, Guastavino was a great vault builder. He possessed the intuition born of the knowledge that the crucial problem in the design of masonry structures is not the resistance of the material but the geometry of the structure. This is the ancient tradition in the calculation of structures. Besides, when he calculated he used the usual, and correct, equilibrium approach employing simple formulae or graphical analysis. In Figure 10, for example, we see the graphical analysis of the great hollow timbrel arches which were to have supported the brick roof of St. John the Divine (this project was not executed; instead, a conventional metallic truss was built).

Designing a masonry dome requires very little: an approximate calculation of the thrusts (the approximate formulas are sufficient), dimensioning of the system of counter-thrust (buttresses, tension ties, or hoops), and a knowledge of the location where tension occurs in order to place supporting diaphragms or fill, which allow for the 'escape' of the forces outside of the surface of the dome. In the case of domes, metal rings (hoops) serve this purpose and vary the direction of the forces. All of this is related strictly to the geometrical form of the vaulting and Guastavino himself affirms:

> *The material of a dome is not only working by compression, but in consequence of its form, it is also working by tension, because* the thrust depends upon the form and not on the material [Guastavino 1893: 75-76; Guastavino's emphasis].

With extraordinary skill, Guastavino employed iron hoops to control the flow of the thrusts within the masonry. He used also many other devices, such as flying buttresses, dwarf vaults and massive cornices, for the same purpose. The study of the constructive sections of Guastavino's domes kept in the Avery Library is fascinating. These designs are evidently the work of a great master of vault construction; see, for example, Figure 11.

However, there is a clear contradiction between both manners of thinking, and the ensuing 'schizophrenia' is manifested in Guastavino's writing and speaking, but not in the constructed work, which is the best proof of Guastavino's mastery.

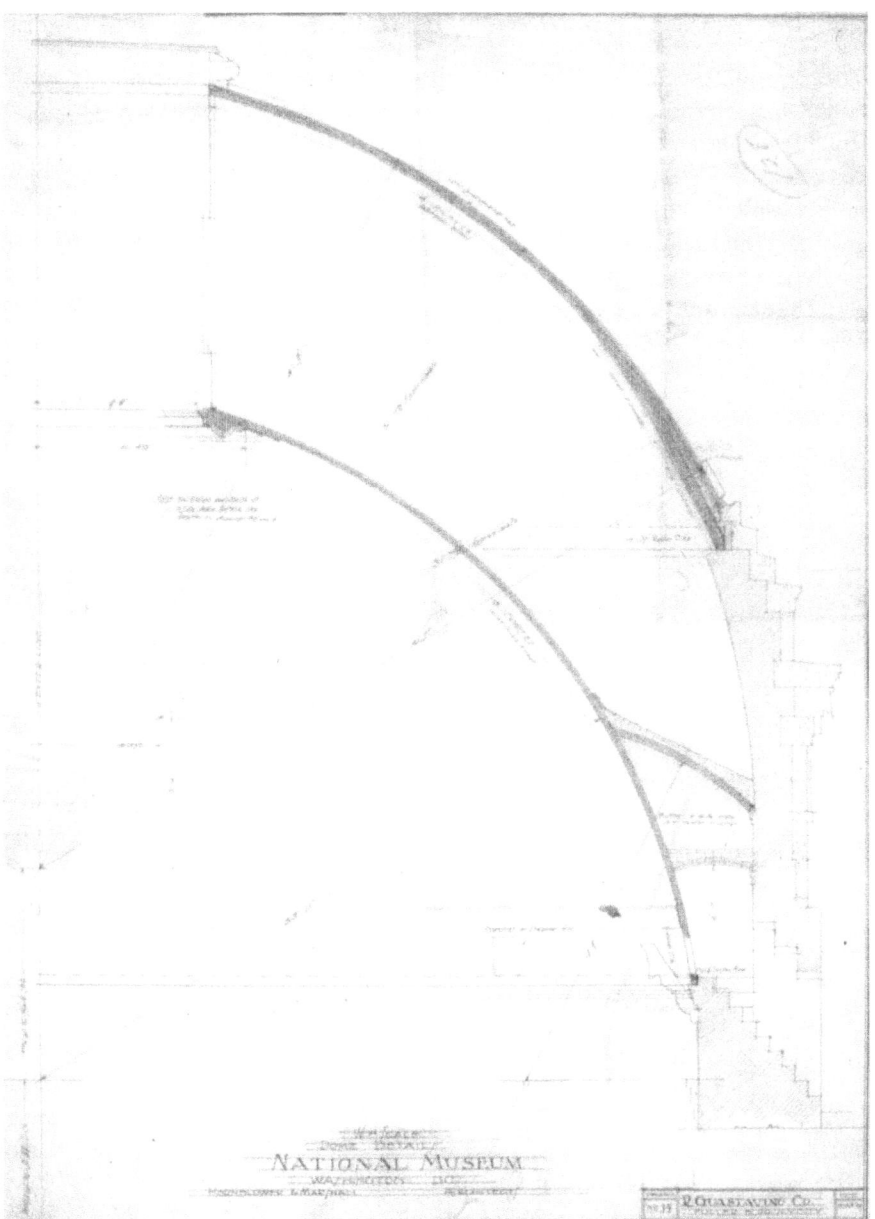

Figure 11. Double timbrel dome in the National Museum of Washington, 1906.
Note the variations of curvature of both shells to avoid tensions and the different devices
to resist the thrust of the domes, including metallic rings, dwarf vaults, flying buttresses
and heavy stone cornices (Avery Library, Columbia University)

Membrane analysis of timbrel vaults: Rafael Guastavino, Jr.

Timbrel domes are very thin shells and the use of membrane analysis to calculate the internal forces seems obvious today for any architect or engineer. Essentially membrane analysis is an equilibrium analysis where all the internal forces are contained within the middle surface of the dome [Heyman 1977]. Simple formulae for membrane analysis of domes of revolution were given already by Rankine in 1858 [Rankine 1864: 265-8]. Schwedler [1866] developed an analytical method for trussed domes that could be extrapolated to thin shells. Eddy [1878] proposed the first graphic method, which was popularised in two articles by Dunn [1904 and 1908]. Eddy's method permits the approximate analysis of domes of revolution of any form (Figure 12).[10]

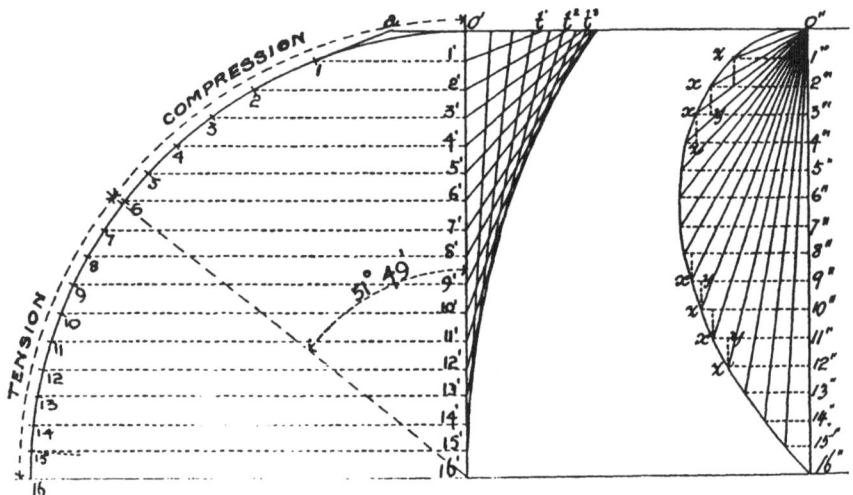

Figure 12. Eddy's graphical method for the membrane analysis of metal or masonry domes [Dunn 1904]

Rafael Guastavino, Jr. (1873-1950) worked with his father in the business from the age of fifteen. He received a "medieval" training, living and working with his father like an apprentice at the feet of a master. In addition, he taught himself art, architecture and structures in his spare time. It is most probable that

---

[10] The history of this method is interesting. Eddy's book was translated into German, *Neue Constructionen aus der graphischen Statik* [Leipzig, 1880]. Föppl [1881] used it without citation. Forty years later, Dischinger [1928] explained it as an analytical graphical method for the calculation of forces in thin shells of any form, again, without citing the origin. From that point on, it appeared in many manuals.

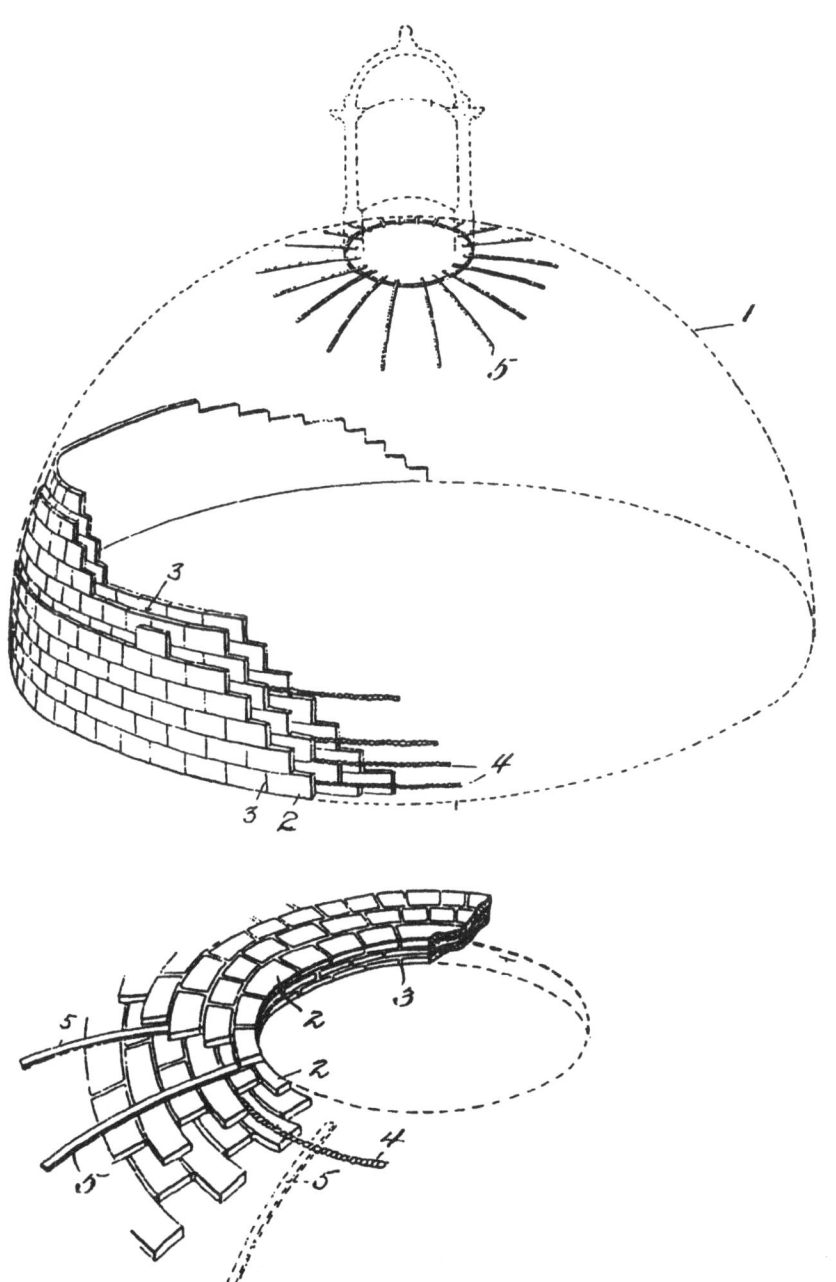

Figure 13. Placement of metal reinforcing in timbrel domes. (Guastavino, Jr., Patent, 1910)

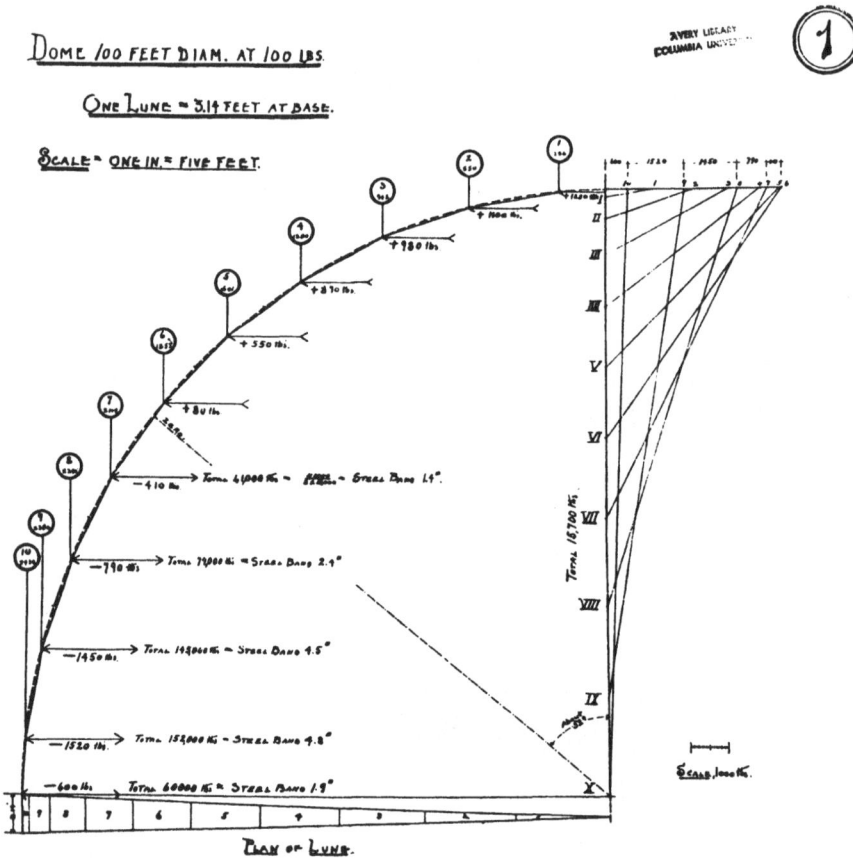

Figure 14. Graphical analysis of a thin dome of with a span of 100 ft: Plan of lune
(Avery Library, Columbia University)

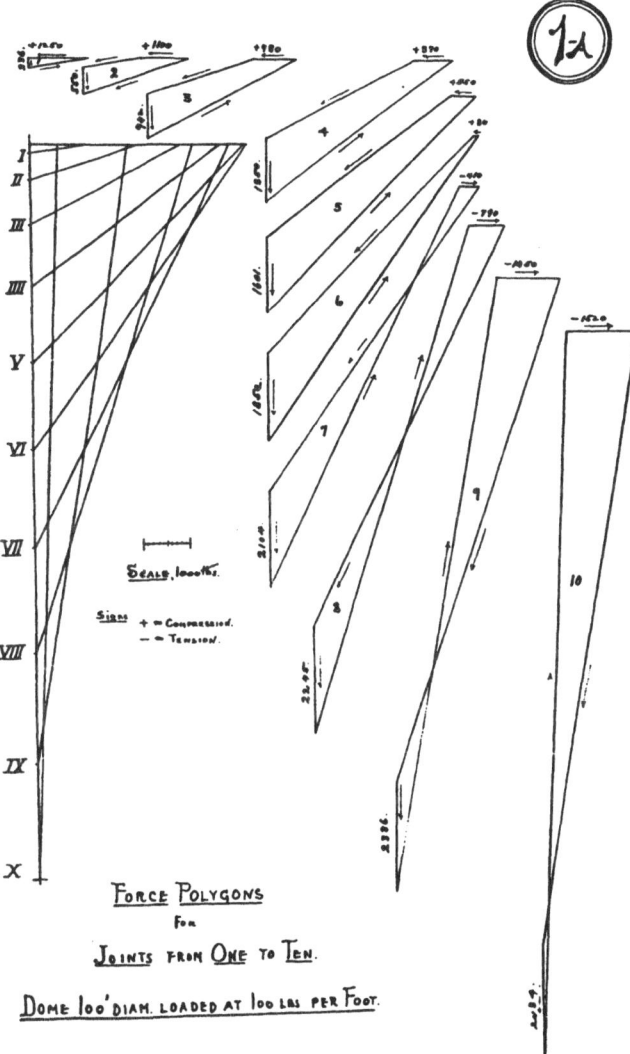

Figure 15. Graphical analysis of a thin dome of with a span of 100 ft: Force polygons
(Avery Library, Columbia University)

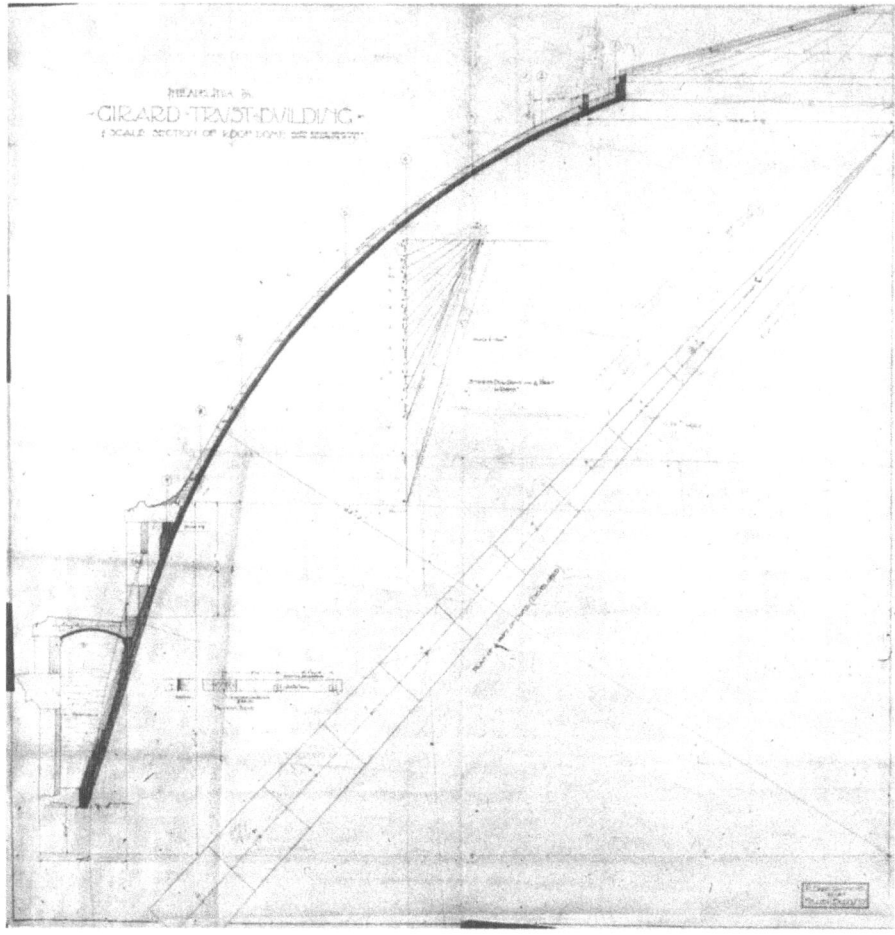

Figure 16. Design of tension-free timbrel dome. Note the change of curvature below the point of zero-stress, the horizontal component of the thrust remaining constant below; see the force polygon. Dome of the Girard Trust Building 1905-1907; 101 ft span (31 m) (Avery Library, Columbia University)

he read Dunn's contribution and decided to apply the method to the analysis of timbrel domes. In particular, he wanted to estimate the tension stresses so that it would possible to calculate and place iron reinforcement. There are two critical places: at the oculus when there is a lantern and at the base (below 52° from the top in a closed hemispherical dome). The method allows one to locate the extension of the tension zones and provide reinforcement easily. In fact, Guastavino, Jr. patented this idea in 1910 (Figure 13).

Guastavino, Jr. made this analysis for many domes, and in particular, for the great temporary dome of St. John the Divine, where metallic reinforcement was placed. In Figures 14 and 15 is shown what appears to be one of the preliminary calculations for a dome with a span of 100 ft. Guastavino, Jr. used the modified version of Dunn [1904]. It is interesting that on the second page, shown in Figure 15, the force polygon is "exploded", in order to understand better the method. It may have been an example for self-study.

Besides, Eddy's observation, republished by Dunn, that from the appearance of tension the thrust remained constant (in a dome constructed with unreinforced masonry), supplied a method for the design of domes without tension. The upper part was a spherical shell, and from the location where tension appeared, the geometry of the lower part of the dome could be traced from the force diagram to give the form of a dome without any tension. Then no reinforcement is needed and the tension hoop rings at the base may easily be calculated. The Guastavinos made extensive use of this discovery in dome design (Figure 16); see also numerous examples in Huerta [2001b: 303-313]. In fact, the approach is better than the complete catenary approach (for example of Gaudí); the dome has a simple geometrical form in the upper part and only deviates from it when it is needed in the lower parts.

Guastavino, Jr. did not publish any papers, but he gave several seminars.[11] In fact, he was profoundly affected by the decline of masonry construction, which he had learned and practised his entire life. To maintain the business, he researched various chromatic possibilities for brick, and above all, he made pioneering research into acoustic materials by collaborating with the leading expert of the period, W.C. Sabine. In the 1930s, he competed with the rise of thin shell construction in concrete, and his interest in this subject is well documented at the Avery Library of Columbia University. Finally, he built a timbrel dome for the Buhl Planetarium of 1938, though such domes were mostly built in reinforced concrete following Dischinger's pioneering work. But this was no longer the age of masonry construction, and the company survived into the

---

[11] In the Guastavino archive of the Avery library, a manuscript is preserved for a magazine article of 1929 and the text of a conference seminar given around 1914.

1960s by developing acoustic materials and building vaults for the last historicist buildings.

Elastic analysis: Domenech, Bayó, Terradas

The table produced by Lanza for Guastavino's *Essay* of 1893 is probably the first evidence of elastic analysis of a timbrel arch. In fact, by the end of the nineteenth century elastic analysis was considered the best approach for masonry arches. The discontinuity and heterogeneity of the masonry, the difficulty in obtaining the elastic constants, the movements during construction, the cracking, etc., were evident, and some engineers were conscious of the dubious character of elastic assumptions applied to masonry arches (see for example [Swain 1927: 425]), but the force of elastic ideas was so great as to overcome any resistance.

Elastic ideas of continuity, tension and bending strength, fit well with Espie's monolithism and Guastavino's cohesion. The only, fundamental, difference is that elastic arches do thrust. The emphasis, then, is in the bending and tension strength of the timbrel vaults. Following Bergós, Gaudí made some calculations to take into account the bending strength of timbrel arches. However, it should be kept in mind that Gaudí published nothing on these matters and that the indirect testimony of Bergós could be biased by his own ideas. A question arises: If Gaudí believed in the bending strength of masonry, why did he use bending-free catenary models? (For the hanging models of Gaudí, see Tomlow [1989].)

It was José Domenech y Estapá who first considered the necessity of taking into account the resistance to bending moments. For Domenech there was no doubt that the only explanation for the success of the thin timbrel vaults came from its capacity to resist bending moments that could cancel the horizontal thrust:

> *The mechanical secret to the construction of these vaults…is not in limiting the calculation of the compressive strength of the materials used, but in taking advantage of the tensile resistance and transverse strength offered by our bricks combined with lime or cement mortar.*
>
> *Utilizing these two strengths, the Catalonian builder could dare to subject his vaults to loads which are unthinkable in others [structures]…always with a small horizontal thrust at the supports, and in some cases reducing the thrust to zero* [Domenech 1900: 38-39].

Once again, the rigid monolithic idea of Espie appears, together with the myth of the absence of thrusts and the cohesive resistance to bending. Later, however, Domenech makes a lucid analysis of the function of timbrel arches, taking as an example the case of a uniform load, in which the line of thrusts is a parabola. He observes that if the directrix of the arch coincides with the line of thrusts (i.e., the arches are exactly parabolic), then there would only be compression, but he leaves this aside. Domenech continues by explaining the method of finding the bending moments and shear forces for a given line of thrust. Finally, he discusses the problem of the position of the line of thrusts, considering the possibility of cracking ("joints of rotation") in an arch (Figure 17).

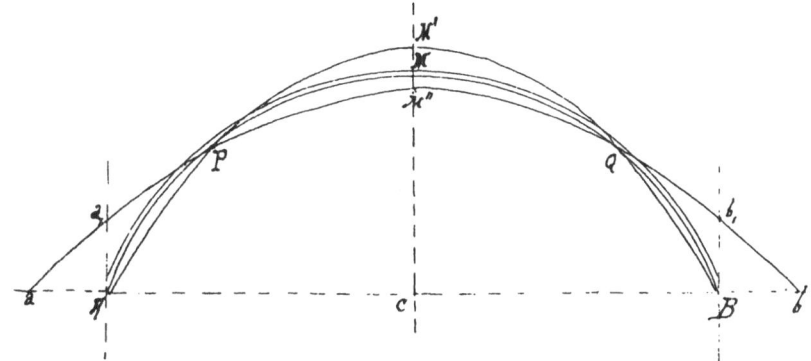

Figure 17. Possible positions of the thrust line in a timbrel arch [Domenech 1900]

Domenech's drawing shows the line of thrust outside the masonry and producing bending moments. In fact, Domenech commits a frequent error: identifying the structure only as the vault, forgetting the fill over the supports and the transverse diaphragms which support the vault. These elements *are also structural* and offer alternative paths for the thrusts to reach the supports. (The supports may be buttresses or masonry walls, a horizontal metal beam supporting ties, etc.) In any case, the situation cannot possibly be maintained over time, due to the low tensile resistance of the masonry, its fragile character, which allows cracking easily and, above all, the inescapable necessity of cracking to adapt to small displacements of the abutments.

Martorell wrote about similar considerations on the bending resistance and the resulting reduction of thrusts:

> *The methods of graphical mechanics used generally, applied to brick arches and in a special way to timbrel arches, give results which are less favourable than the corresponding reality . . . The cohesion, the rigidity of timbrel vaults, greatly lowers the thrust and at the same time, allows them to be built in implausible forms, as if they were metallic shells* [Martorell 1910: 143].

Martorell alludes to the various positions of the line of thrust, and implicitly, to the appearance of bending moments, highlighting the necessity for tests allowing the calculation of the "coefficients used in calculations to evaluate the bending resistance and the transverse forces in the timbrel vaults".

Jaime Bayó [1910] is the first in Spain to propose a proper elastic analysis for timbrel vaults. In his article, he equates them with metal arches (two-hinged), criticising the traditional methods for the calculation of voussoir arches:

> To calculate this [timbrel] vault... by determining the line of thrust of a voussoir arch, is born of an error, which supposes that it only works in compression. This is not the case, as it works also in tension, being like any metal surface it can support bending [Bayó 1910: 165].

For Bayó the timbrel vaults thrust, but this thrust corresponds to that of a two-hinged, elastic arch. He tries to find what he calls "the funicular of the elastic forces", that is, the line of thrust that is in equilibrium with the loads, and also complies with the compatibility conditions of elastic deformations. Bayó gives the formulas with the usual integrals, and later, he explains a graphic solution method, applying it first to symmetrical arches of constant or variable thickness, and then to asymmetrical arches. He also explains how to calculate the tensile and compressive stresses, and cites the experimental tests by Guastavino as a current reference to consider the values for allowable stress: 1.5 N/mm^2 in compression, 0.4-0.5 N/mm^2 in tension, and 0.6 N/mm^2 in shear. The article finishes with some considerations on the design of timbrel vaults, in which he recommends adjusting the thickness (number of brick layers) according to the bending stresses. He observes that in the case of the flat vaults, the thrust can be calculated as if it were made of voussoirs, working only in compression, but if they are higher it is precise to adapt the form to the line of thrust. He finishes the article proposing a method to design timbrel vaults of any shape:

> If it is desired to construct equilibrated vaults or of equal resistance, that respond to the design suggested by the imagination of an artist, one should proceed as shown in the figure... ...after determining the funicular of the elastic forces, the thickness of the vault is given in relation to the value of the bending moments [Bayó 1910: 184] (Figure 18).

We have no evidence that Bayó or any other architect ever constructed a timbrel vault (or any other masonry structure), with this form and thickness, but the drawing clearly shows the strong belief in tensile resistance, in elastic calculations, and in the cohesive properties of timbrel vaults.

Finally, it must be noted that the calculation of timbrel arches corresponds to the most elementary example of timbrel vault construction. Bayó does not mention the calculation of the more complex and most common forms, such as the crossing vault, barrel vault, ribbed vault and domes. For the calculation of internal forces in these cases, the only viable approximation was the equilibrium calculations using graphic statics or hanging models and this was the common practice.

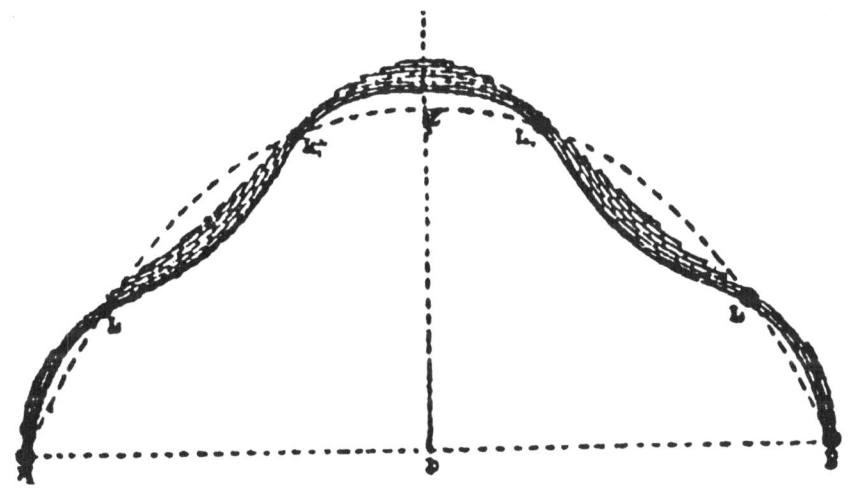

Figure 18. Timbrel vault of a peculiar form (impossible to construct in reality) where the thickness is prescribed according to the magnitude of the bending moments resulting from an elastic analysis [Bayó 1910]

Consolidation of the elastic-cohesive focus: vaults 'impossible to calculate'

The cohesive ideas formulated first by Espie, taken up, expanded and spread by Guastavino, and later adopted by elastic theory, were converted into a dogma. In a book of 1910 on the philosophy of structures, Cardellach frames the timbrel vaults under cohesive construction, highlights the capacity to resist bending and, like Bayó, insists on the infinite variety of forms that can be constructed.

Esteve Terradas, the great Spanish engineer and mathematician, was the first to try an elastic analysis of a more complex timbrel vault: the vault of a staircase. The contribution of Terradas has been analysed in detail by Rosell and Serrá [1987]. In this context it must be stressed that Terradas's study had its origin in the challenge of Puig y Cadafalch, in 1919, to solve the problem of the calculation for timbrel vaults. As Rosell and Serrá write, "the habitual vaults, constructed by bricklayers 'from feeling', were considered to be impossible to calculate" [Rosell and Serrà 1987: 24]. Terradas collected his sketches, notes, and calculations in a small book called *Libreta de la volta.*[12]

Terradas tried to make an elastic analysis of the vaults and he examined well-known elastic problems, in particular that of buckling. He failed in his attempt. The planning of the elastic equilibrium equations for a spatial structure such as a vaulted staircase is very complex. The failure of Terradas served to reinforce the idea of the impossibility of calculating the forces in timbrel vaults.

Later, Josep Goday, in his 1934 discourse before the Acadèmia Catalana de Belles Arts de Sant Jordi, made a historical survey of the calculation of timbrel vaulting. He accepts the cohesive ideas of Guastavino and agrees with Bayó and Terradas that the only correct focus is to consider the vaults as thin, continuous, elastic shells, within the context of elastic theory. At the end of his article, Goday briefly discusses membrane analysis, but does not seem to appreciate that it is an equilibrium method which does not consider the material properties. (The membrane theory, dating from the second half of the nineteenth century, was popularised in Europe in the 1930s, principally through the theoretical and practical work of the German engineer Franz Dischinger, using it to design thin shells of reinforced concrete.)

These confirmed the idea that timbrel vaults could only be calculated as elastic, and if this task presented insurmountable difficulties, then the vaults were impossible to calculate. Eduardo Torroja, the great Spanish engineer and builder of thin concrete shells, repeated this opinion, writing, "so marvellous in its constructions, that modern theories have difficulty explaining and measuring its phenomenal resistance, so brilliantly intuited by builders of the past" [Torroja 1956]. Bassegoda, in his numerous contributions on timbrel vaulting [Huerta et al. 2001], expressed similar opinions, and more recently, Professor J. L. González [1999] considered it necessary to make a load test on a timbrel vault stair to reliably estimate its strength.

Calculations in Practice

As Rankine noted accurately in the Introductory Essay to his book on *Applied Mechanics*, if the question of theoretical science is "what are we to think?", the question in practical science is "what are we to do?" [Rankine 1858: 10]. An insufficient theory, real or imagined, has never deterred the builders who have

---

[12] Professor Rosell provided me with a photocopy of this book and volunteered his time to speak with me on various aspects of Terrada's work. For all of this, I am extremely grateful to him.

applied the available tools of the time period. Thus, while the engineer-scientists argued over the impossibility of calculating the forces in timbrel vaulting, the builders continued constructing and the architects or engineers made simple calculations to determine the dimensions of the principal elements: the thickness of the vaults, and the sizing of the systems to resist the thrust.

The builders' clear belief in the thrust of timbrel vaults can be demonstrated by their use of systems to counter thrust, which are always present. As we have seen, traditional rules for buttress design for timbrel vaults have existed. The French engineers of the nineteenth century made equilibrium calculations, as did the Guastavinos, though the hypotheses behind the formulas were in direct opposition to the cohesive theory. Luis Moya [1957], the last great builder of timbrel vaults, acknowledged the insufficiency of calculations owing to the lack of data on the elastic constants of timbrel vaults, but later he made, or directed another to make, equilibrium calculations based on the line of thrust to design and build his astonishing vaults.

Bosch [1947] spoke in favour of membrane analysis but for practical cases proposed an ingenious system (inspired, no doubt, by nineteenth century manuals on the theory of vaulting) to calculate the thrust of timbrel vaulting, by cutting the vault into a series of arches. He imagined the existence of virtual crossing ribs which support a series of parallel arches between the ribs. Again, this is an equilibrium method that seeks to find one possible state of compression within the masonry.

Bergós [1936; 1953; 1965] dedicated several decades to studying the mechanical properties of masonry walls and timbrel vaults. He tested timbrel arches of various sizes (up to 3.2 m in span), trying to justify the application of elastic theory. But in the examples of practical calculation that appear in his books he uses graphic methods of thrust lines, that is, equilibrium methods.

Angel Pereda Bacigalupi [1951] published one of the last books on the calculation of timbrel vaulting. Like Bayó, he supposed two-hinged arches on rigid supports, and calculated them with the usual equations for elastic arches. Vaults were often built with tension ties to take the thrust, but the deformation of this tie was not considered in the calculation, even though it would lead to significant bending moments. In fact, Pereda realized that an elastic calculation could not pretend to account for the flexural resistance of timbrel vaulting. He explicitly looks for the thickness so that the line of elastic thrusts is contained within the middle third of the section. To accomplish this, Pereda lowered the admissible tensile stress, demonstrating a better knowledge of the material properties than his earlier predecessors working with elastic calculations.

The use of Finite Element Methods

Today, the finite element method (FEM) has been applied to the analysis of timbrel vault structures. Gulli [1993;1994; 1995] made tests on barrel vaults and later, implemented finite element methods to carry out elastic calculations. The finite element method, like traditional elastic calculations, considers the masonry as a continuum with certain elastic properties, which require assumptions about the support conditions. These assumptions about the supports and the material, together with static equilibrium, form a system of equations that give a unique solution. This focus presents various problems. In the first place, the system of equations is extremely sensitive to small variations in the support conditions. For example, a small settlement or rotation of one of the supports, imperceptible to the eye, will give a large variation in the system of internal forces (and the analyst can use a FEM program to verify this point). In the second place, timbrel construction is far from a continuum and is frequently cracked. The use of FEM programs allow a non-linear analysis, which improves the model, but is still highly sensitive to changes in the support conditions, the load history of the structure, the formation of cracks in unexpected locations, etc. In summary, the results from an elastic analysis or the FEM have little significance, and are of no assistance in understanding the structural behaviour of the timbrel vault or masonry structure in question.

Conclusion : The timbrel vault as a masonry vault

Timbrel vaults are masonry vaults, with a good strength in compression, a low tensile strength and the possibility of cracking, forming 'hinges', due to the impossibility of sliding. In fact, in a hyperstatic masonry structure the formation of cracks is inevitable, and traditional timbrel vaulting shows the same pathologies of stone or brick vaulting.

It is true that the tensile resistance also allows a certain amount of bending resistance. For example, a mason can walk on a thin timbrel arch. To explain the resistance to higher loads or over longer periods of time, one must look to the other resisting elements. Thin walls, or lateral diaphragms, as well as a solid fill, which forms the base for a floor, are actually part of the resisting structure and help to resist moving loads. The principle is always the same: to give a sound escape route for the forces when necessary, or to load the structures so that the line of forces is always contained within the masonry.

The possibility of perforating a vault without collapse, which has been cited since Espie as a characteristic of the cohesive structure, is also true of other masonry vaults. Every so often a pinnacle falls from a buttress and perforates a Gothic crossing vault without causing collapse.

The 'cohesive' character is not relevant from the structural point of view, but is important from the constructive point of view. It allows for the construction without centering, using only light auxiliary elements to control the form. In addition, timbrel vaults present some resistance to bending, which permits the passage of light loads during construction, making the process even easier.

Summing up, the fundamental statements about the 'masonry' material (high compressive strength, low tensile strength and no sliding) apply also to timbrel vaults. Professor Heyman has systematised these properties to include masonry structures within the more general framework of Limit Analysis. From his first article in 1966 until today he has clearly illustrated the theory by applying it to basic structural elements: buttresses, domes, crossing vaults, spires, towers, bridges, etc. (The articles have been compiled in Heyman [1995], as well as an overview of his work in [1997].)

Within the frame of Limit Analysis, the Safe Theorem validates the approach of equilibrium: if we can find an equilibrium solution for the masonry structure with the material working in compression, then the structure is safe. The power of the theorem lies in that we may 'choose' the equilibrium solution. If the analyst can find a situation of equilibrium in compression. the structure will be able to as well. In fact, the analyst will study only some of the infinitely many equilibrium states possible in a hyperstatic structure.

The equilibrium analysis of the old theory of vaults is therefore perfectly correct, and lies within the scope of Limit Analysis [Huerta 2001].[13] The simplified formulas of Guastavino, the graphic analysis, Gaudí's use of hanging models, and membrane analysis of compression states by Guastavino, Jr., are all correct. Elastic analysis 'in compression', like that of Pereda cited above, are also correct. To state it more clearly, they are 'safe': a structure designed on the basis of them will not fall and the same methods can be used to measure their safety. In fact, there could be no other conclusion for structures which have survived for more than a century, and this experimental demonstration is conclusive.

Even more, the traditional proportional rules for the design of vaults and buttresses (like those of Fray Lorenzo) are also essentially correct [Huerta 1990; 1999]. The problem of the safety of a masonry vault, whether of stone, brick, mass concrete or a timbrel vault, is a problem in the geometrical form of the structure. The stable forms contain lines of thrust in equilibrium with the applied loads. The traditional rules codify these forms and their use is rational

---

[13] The approach of equilibrium has been recently applied to the calculation of the thrust of timbrel groined vaults by Fortea and López [1998].

and correct. (Of course these rules are particular for each type of structure: a Gothic buttress would not support the thrust of a Roman vault.)

Acknowledgment

I would like to thank John Ochsendorf for his kind assistance in preparing the English manuscript of this article.

Bibliography

ARAGUAS, PHILIPPE. 1999. Voûte a la rousillon. *Butlletí de la Reial Academia Catalana de Belles Arts Sant Jordi*, vol. 13: 173–185.

BAILS, BENITO. 1796. *Elementos de Matemáticas. Tomo IX. Parte I. Que trata de la Arquitectura Civil.* Madrid: Imprenta de la Viuda de Joachim Ibarra (facs. ed. Murcia: C. O. de Aparejadores y Arquitectos Técnicos,1983).

BANNISTER, T.C. 1968. The Roussillon Vault. The Apotheosis of a 'Folk' Construction. *Journal of the Society of Architectural Historians,* 27:163–75.

BASSEGODA NONELL, JOAN. 2001. La obra arquitectónica de Rafael Guastavino en Cataluña (1866-1881). In *Las bóvedas de Guastavino en América.* S. Huerta (ed.). Madrid: Instituto Juan de Herrera, CEHOPU: 373-393.

BAYÓ, JAIME. 1910. La bóveda tabicada. *Anuario de la Asociación de Arquitectos de Cataluña*:157–84.

BÉLIDOR, B.F. 1729. *La science des ingénieurs dans la conduite des travaux de fortification et architecture civile.* Paris.

BENVENUTO, EDOARDO. 1991. *An Introduction to the History of Structural Mechanics. Part II: Vaulted Structures and Elastic Systems.* New York/Berlin: Springer Verlag.

BERGÓS MASSÓ, JUAN. 1936. *Formulario técnico de construcciones.* Barcelona: Bosch.

———. 1953. *Materiales y elementos de construcción.Estudio experimental.* Barcelona: Bosch.

———. 1965. *Tabicados huecos.* Barcelona: Colegio de Arquitectos de Cataluña y Baleares.

BLONDEL, J.F. 1771–77. *Cours d'Architecture, ou Traité de la décoration, distribution et construction des bâtiments... continué par M. Patte.* Paris: Chez la Veuve Desaint.

BOSCH REITG, IGNACIO. 1949. La bóveda vaida tabicada. *Revista Nacional de Arquitectura*:185–99.

CARDELLACH, FELIX. 1970. *Filosofía de las Estructuras.* Barcelona: Editores Técnicos Asociados. (1st. ed. 1910.)

CHOISY, AUGUSTE. 1883. *L'Art de Bâtir chez les Byzantines.* Paris.

COLLINS, GEORGE R. 1968. The Transfer of Thin Masonry Vaulting from Spain to America. *Journal of the Society of Architectural Historians,* vol. 27:

176–201. (Spanish translation in *Las bóvedas de Guastavino en América*. S. Huerta (ed.). Madrid: Instituto Juan de Herrera, CEHOPU: 19-45.)

DEJARDIN, M. 1860. *Routine de l'etablissement des voutes...* 2nd ed. Paris: Dalmont et Dunod.

DELHUMEAU, G. 1999. *L'invention du béton armé: Hennebique 1890–1914.* Paris: Éditions Norma.

DISCHINGER, FRANZ. 1928. *Schalen und Rippenkuppeln.* (4a ed. *Handbuch der Eisenbetonbau.* VI Band, Zweiter Teil., F. von Emperger ed.). Berlín: Wilhelm Ernst und Sohn.

D'OLIVIER. 1837. Relatif à la construction des voûtes en briques posées de plat, suivi du recherches expérimentales sur la poussée de ces sortes des voûtes. *Annales des Ponts et Chaussées, 1er série*, 292–309, Pl. 129.

DOMENECH Y ESTAPA, JOSE. 1900. La fábrica de ladrillo en la construcción catalana. *Anuario de la Asociación de Arquitectos de Cataluña:* 37–48.

DUNN, W. 1904. Notes on the Stresses in Framed Spires and Domes. *Journal of the Royal Institute of British Architects, Third series,* vol. 11 (Nov. 1903 - Oct. 1904): 401–412.

———. 1908. The Principles of Dome Construction. *Architectural Review,* vol. 23: 63-73; 108–112.

EDDY, HENRY T. 1878. *Researches in Graphical Statics.* New York: Van Nostrand.

ESPIE, COMTE D'. 1754. *Manière de rendre toutes sortes d'édifices incombustibles...* Paris: Duchesne.

FONTAINE, H. 1865. Expériences faites sur la stabilité des Voûtes en briques. *Nouvelles Annales de la Construction,* vol. 11: 149–159, Plate 45.

FÖPPL, AUGUST. 1881. *Theorie der Gewölbe.* Leipzig: Felix.

FORNÉS Y GURREA, MANUEL. 1841. *Observaciones sobre la práctica del arte de edificar.* Valencia: Imprenta de Cabrerizo. (facs. ed. Valencia: Librería París-Valencia, 1993.)

FORNES Y GURREA, MANUEL. 1846. *Álbum de proyectos originales de arquitectura, acompañado de lecciones explicativas.* Valencia: Imprenta de D. Mariano Cabrerizo. (facs. ed. Madrid: Ediciones Poniente, 1982.)

———. 1857. *Observaciones sobre el arte de edificar.* Valencia: Imprenta de D. Mariano Cabrerizo. (facs. ed. Madrid: Ediciones Poniente, 1982.)

FORTEA LUNA, MANUEL and VICENTE LÓPEZ BERNAL. 1998. *Bóvedas extremeñas. Proceso constructivo y análisis estructural de bóvedas de arista.* Badajoz: Colegio Oficial de Arquitectos de Extremadura.

FRÉZIER, A.F. 1754-69. *La théorie et la pratique de la coupe de pierres et des bois... ou traité de stéréotomie à l'usage de l'architecture.* Strasbourg/Paris: Charles-Antoine Jombert. (1st. ed. 1737-1739.)

GARCÍA BERRUGUILLA, JUAN. 1747. *Verdadera práctica de las resoluciones de la Geometría...* Madrid: Imprenta de Lorenzo Francisco Mojados. (facs. ed. Murcia: C. O. de Aparejadores y Arquitectos Técnicos, 1979.)

GER Y LOBEZ, FLORENCIO. 1915. *Manual de construcción civil.* 2nd ed. Badajoz: La Minerva Extremeña, (1st ed. Badajoz, 1869.)

GODAY, JOSEP. 1934. *Estudi històric i mètodes de càlcul de les voltes de maó de pla* Barcelona: Acadèmia Catalana de Belles Arts de Sant Jordi.

GONZÁLEZ MORENO-NAVARRO, JOSÉ LUIS. 1999. La bóveda tabicada. Su historia y su futuro. Pp. 237-259 in *Teoría e historia de la restauración.* Vol. 1. Madrid: Munilla-Llería.

GUASTAVINO, SR., RAFAEL. 1893. *Essay on the Theory and History of Cohesive Construction, applied especially to the timbrel vault.* Boston: Ticknor and Company. (1st. ed. 1892.)

———. 1896–1904. *Prolegomenos on the function of masonry in modern architectural structures.* New York: Record & Guide Press.

GULLI, RICCARDO.1993. Le volte in folio portanti: Tecnica costruttiva ed impiego nell'edilizia storica e moderna. In *Atti del I Convegno Nazionale Manutenzione e Recupero nella Città Storica, ARCO,* 595B604. Rome.

———. 1993. Il sistema tabicado. Una tecnica tradizionale per il recupero. En *Atti del Convegno Internazionale: Il recupero degli edifici antichi, manualistica e nuove tecnologie,* 198B208. Naples.

———. 1994. Una ipotesi di intervento conservativo per il recupero delle volte in folio portanti. En *Atti del Convegno di Studi: La ricerca del recupero edilizio, Ancona,* 51B62. Bologna.

———. 2001. Arte y técnica de la construcción tabicada. In *Las bóvedas de Guastavino en América.* S. Huerta (ed.). Madrid: Instituto Juan de Herrera, CEHOPU: 59-85.

GULLI, RICCARDO and GIOVANNI MOCHI. 1995. *Bóvedas tabicadas: Architettura e costruzione.* Rome: CDP Editrice.

Heyman, Jacques. 1977. *Equilibrium of shell structures.* Oxford: Oxford University Press.

———. 1982. *The masonry arch.* Chichester: Ellis Horwood.

———. 1995. *Arches, vaults, and buttresses: Masonry structures and their engineering. Collection of essays.* London: Variorum.

———. 1997. *The stone skeleton: Structural engineering of masonry architecture.* Cambridge: Cambridge University Press.

———. 1999. *The science of structural engineering,* London: Imperial College Press.

HUERTA, SANTIAGO. 1990. *Diseño estructural de arcos, bóvedas y cúpulas en España, ca. 1500–ca. 1800.* Ph.D. Diss. Universidad Politécnica de Madrid, Escuela Técnica Superior de Arquitectura.

———. 1996. La teoría del arco de fábrica: desarrollo histórico. *Obra Pública,*

n. 38: 18–29.

———. 1999. The medieval 'scientia' of structures: the rules of Rodrigo Gil de Hontañón. *Omaggio a Edoardo Benvenuto,* Genoa 29-30 November, 1 December 1999.

———. 2001a. Mechanics of masonry vaults: the equilibrium approach. *Structural analysis of historical constructions III. Possibilities of numerical and experimental techniques.* P. B. Lourenço and P. Roca, eds. Guimaraes: Universidade do Minho.

——— (ed). 2001b. *Las bóvedas de Guastavino en América.* Exhibit book-catalogue. Madrid: Instituto Juan de Herrera, CEHOPU.

———. 2001c. La mecánica de las bóvedas tabicadas en su contexto histórico: la aportación de los Guastavino. In *Las bóvedas de Guastavino en América.* S. Huerta, ed., Madrid: Instituto Juan de Herrera, CEHOPU: 87-112.

HUERTA, SANTIAGO, GEMA LÓPEZ and ESTHER REDONDO. 2001. Bibliografía seleccionada y comentada sobre Guastavino y la construcción tabicada. In *Las bóvedas de Guastavino en América.* S. Huerta (ed.). Madrid: Instituto Juan de Herrera, CEHOPU: 373-393.

KOESTLER, ARTHUR. 1964. *The act of creation.* New York. Macmillan.

LANZA, GAETANO. 1891. *Applied mechanics.* New York: John Wiley and Sons. (1st ed. 1885)

LEMMA, MASSIMO. 1996. *Dei tetti ammattonati. Nuova edizione critica del tratatto scritto da Felix François d'Espie (1754).* Venice: Il Cardo.

LEMMONIER, M. HENRY. 1920. *Procès-verbaux de l'Académie Royale d'Architecture, 1671–1793. Tome VI: 1744–1758.* Paris: Édouard Champion.

MARIAS, FERNANDO. 1991. Piedra y ladrillo en la arquitectura española del siglo XVI. Pp. 71-83 in *Les chantiers de la Renaissance,* J. Guillaume ed. París: Picard.

MARTORELL, JERÓNIMO. 1910. Estructuras de ladrillo y hierro atirantado en la arquitectura catalana moderna. *Anuario de la Asociación de Arquitectos de Cataluña:* 119–146.

MOCHI, GIOVANNI. 2001. Elementos para una historia de la construcción tabicda. In *Las bóvedas de Guastavino en América.* S. Huerta (ed.). Madrid: Instituto Juan de Herrera, CEHOPU: 113-146.

MOYA BLANCO, LUIS. 1957. *Bóvedas Tabicadas.* Madrid: Dirección General de Arquitectura. (facs. ed. Madrid: C. O. de Arquitectos, 1993.)

NEUMANN, DIETRICH. The Guastavino system in context: History and dissemination of a revolutionary vaulting method. *APT (Association of Preservation Technology) Bulletin* 30, 4 (1999): 7–13. (Spanish translation in *Las bóvedas de Guastavino en América.* S. Huerta (ed.). Madrid: Instituto Juan de Herrera, CEHOPU: 147- 154.)

NEWLON, HOWARD ed. 1976. *A Selection of Historic American Papers on Concrete, 1876-1926.* Detroit: American Concrete Institute.

PARKS, JANET and ALAN G. NEUMANN eds. 1996. *The Old world builds the New: The Guastavino Company and the technology of the catalan vault, 1885-1962.* Exhibit catalogue. New York: Avery Architectural Library and the Miriam and Ira D. Wallach Art Gallery, Columbia University.

PARKS, JANET. 2001. Génesis del *Ensayo sobre la construcción cohesiva* de Rafael Guastavino. In *Las bóvedas de Guastavino en América*, S. Huerta ed. Madrid: Instituto Juan de Herrera, CEHOPU : 173-175.

PEREDA BACIGALUPI, ANGEL. 1951. *Bóvedas tabicadas. Cálculo y ejemplos resueltos.* Santander: Editorial Cantabria.

PLO Y CAMÍN, ANTONIO. 1767. *El Arquitecto práctico, civil, militar y Agrimenso...* Madrid: Imprenta de Pantaleón Aznar. (facs. ed. Valencia: Librería París-Valencia, 1995.)

RAMAZOTTI, LUIGI. 2001. La cúpula para San Juan el Divino de Nueva York de Rafael Guastavino. In *Las bóvedas de Guastavino en América*. S. Huerta (ed.). Madrid: Instituto Juan de Herrera, CEHOPU: 187-200.

RANKINE, W. J. M. 1864. *A Manual of Applied Mechanics.* 3rd ed. London: Charles Griffin. (1st ed. 1858.)

RIEGER, P. CHRISTINO. 1763. *Elementos de toda la architectura civi...* Madrid: Joachim Ibarra.

RONDELET, JEAN. 1834-48. *Traité théorique et pratique de l'art de bâtir.* Paris: Chez Firmin Didot. (1st ed. Paris: 1802.)

ROSELL, JAUME and ISABEL SERRA. 1987. Estudis d'Esteve Terradas sobre la volta de maó de pla. Pp. 23-33 in *Cinquanta anys de ciència i tècnica a Catalunya,* Barcelona: Institut dÉstudis Catalans.

ROSELL, JAUME. 2001. Rafael Guastavino Moreno: Ingenio en la arquitectura del s. XIX. In *Las bóvedas de Guastavino en América*. S. Huerta (ed.). Madrid: Instituto Juan de Herrera, CEHOPU: 201-215.

ROSENTHAL, E. E. 1988. *El palacio de Carlos V en Granada.* Madrid: Alianza Forma.

SAN NICOLÁS, FRAY LORENZO DE. 1639. *Arte y Uso de Architectura. Primera parte.* Madrid: s.i. (facs. ed. Madrid: Albatros, 1989)

SCHWEDLER, J. W. 1866. Die Konstruktion der Kuppeldächer. *Zeitschrift für Bauwesen,* vol. 16: 7–34, lám. 10-14.

SOTOMAYOR, JOAQUIN DE. 1776. *Modo de hacer incombustibles los edificios sin aumentar el coste de la construcción. Extractado del que escribió en francés el Conde de Espié.* Madrid: Oficina de Pantaleón Aznar.

SWAIN, GEORGE F. 1927. *Structural Engineering. Stresses, graphical statics and masonry.* New York: McGraw-Hill.

TARRAGÓ, SALVADOR. 2001. Las variaciones históricas de la bóveda tabicada. In *Las bóvedas de Guastavino en América*. S. Huerta (ed.). Madrid: Instituto Juan de Herrera, CEHOPU: 217-240.

TOMLOW, JOS. 1989. *Das Modell. Antoni Gaudis Hängemodell und seine Rekonstruktion. Neue Erkenntnisse zum Entwurf für die Kirche der Colonia Güell.* Stuttgart: Institut für leichte Flächentragwerke, University of Stuttgart.

TOMLOW, JOS. 2001. La bóveda tabicada a la catalana y el nacimiento de la "cerámica armada" en Uruguay. In *Las bóvedas de Guastavino en América.* S. Huerta (ed.). Madrid: Instituto Juan de Herrera, CEHOPU: 241-251.

TORROJA, EDUARDO. 1956. *Razón y ser de los tipos estructurales.* Madrid: Instituto Eduardo Torroja de la Construcción y del Cemento.

TOURTAY, C. 1885. Sur l'influence des joints dans la résistance à l'écrasement des maçonneries de pierres de taille. *Annales des Ponts et Chaussées,* vol. 2: 582-592.

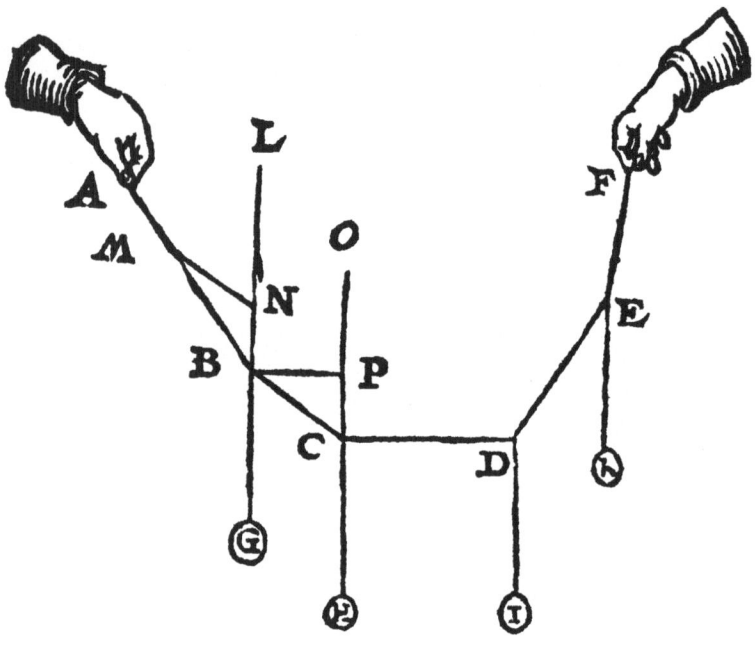

One of Simon Stevin's first figures for the parallelogram law

The Use of a Particular Form of the Parallelogram Law of Forces for the Building of Vaults (1650-1750)

Patricia Radelet-de Grave[1]

One of the principal, if not the only, interests in the history of mechanics of Clifford Truesdell as well as of Edoardo Benvenuto, was to understand how people discovered fundamental notions, laws and principles. Reading those texts on vaults once more, I realised that the history of the study of stability of a vault under its own weight during the period 1650-1750 was important for the discovery of other fundamental notions, so I decided to enlarge the aim of the present paper in order to account for it. I'll try to answer the question of what the study of that problem brought to the knowledge of fundamental mechanical notions.

Introduction

When I told Clifford Truesdell that I was working on the history of the parallelogram law of forces, he asked me if I knew Edoardo Benvenuto, who had written an article on that argument. That is the reason why, some years later, I met Edoardo Benvenuto. Therefore, at the conference in Genoa 30 November-1 December 2001 in honour of Benvenuto and Truesdell, as a record of that first meeting, I wanted to speak of the parallelogram law of forces in the context of the study of the stability of vaults. But as we all know, one of the principal, if not the only, interests in the history of mechanics of Clifford Truesdell as well as of Edoardo Benvenuto, was to understand how people discovered fundamental notions, laws and principles. Reading those texts on vaults once more, I realised that the history of the study of stability of a vault under its own weight during the period 1650-1750 was important for the discovery of other fundamental notions, so I decided to enlarge the aim of the present paper in order to account for it. I'll try to answer the question of what the study of that problem brought to the knowledge of fundamental mechanical notions.

As mentioned, the story will take place during a period from 1650 to1750. I'll tell that story from my point of view using original texts but also referring to the work of Truesdell and Benvenuto. I'll use principally the two books that defined our colloquium and two other books, one by Truesdell, *Essays in the History of Mechanics* [1968], and one by Benvenuto, *La scienza delle costruzioni ed il suo sviluppo storico* [1981].

---

[1] Institut de Physique Théorique (FYMA), Catholic University of Louvain, Chemin de Cyclotron 2, B-1348 Louvain-la-Neuve, BELGIUM

But I should explain first the particular problem of which I intend to study the development because the problem of the stability of an arch has many components. The one I want to concentrate upon is which is the best form to give to an arch in order for it to be stable under its own weight. But the correct definition of the problem is, as always, an essential part of the difficulties scientists will have to overcome.

Analogy between vault and catenary

To understand the history of the stability of vaults and, more precisely, of the best form to give to an arch, it is essential to know that in studying that argument, we are dealing with the history of techniques and that empirical knowledge existed since antiquity, in all civilisations. We can still see it in some important monument. On the other hand, theorisation of that subject arose rather late. In fact the first who tried to theorise the problem was Philippe de la Hire in the *Traité de mécanique* [1695], that is to say, eleven years after Leibniz invented differential calculus.

It is also important to know that the history of the theorisation of the form of the most stable vault is linked to the history of the catenary, the form assumed by a chain hanging freely, and that that history is linked to the birth of the calculus (Figure 1). If, omitting Galileo's question about the catenary and his attempt to solve it by the composition of movements, which lead to the erroneous answer, the parabola, we turn directly to Jacob Bernoulli who posed that question in May 1690, first to Leibniz, then to the scientific community, in order to develop Leibniz's differential calculus.

The stories of the catenary and of the vault were first linked empirically. We can find some ancient vaults that are inverted catenaries. Then by an affirmation of Hooke, who in 1675, at the end of an article famous for giving Hooke's law for the spring, gave some *decimate of the centesme,* N°2 gives an anagram that, when deciphered, reads "Ut pendet continuum flexible line, so but inverted will stand the rigid arch" [Hooke 1675]. Regarding this, Truesdell adds between brackets,

> *While none of the available papers of Hooke reveals how he reached this conclusion, there is no reason to doubt that he had sufficient mastery of statics to show that an arch of infinitely small stones in order to exert purely tangential thrust should be formed like an inverted catenary subject to inverted loads. Thus the problem of the catenary and the arch are reduced to one, but neither is solved* [Truesdell 1960: 57].

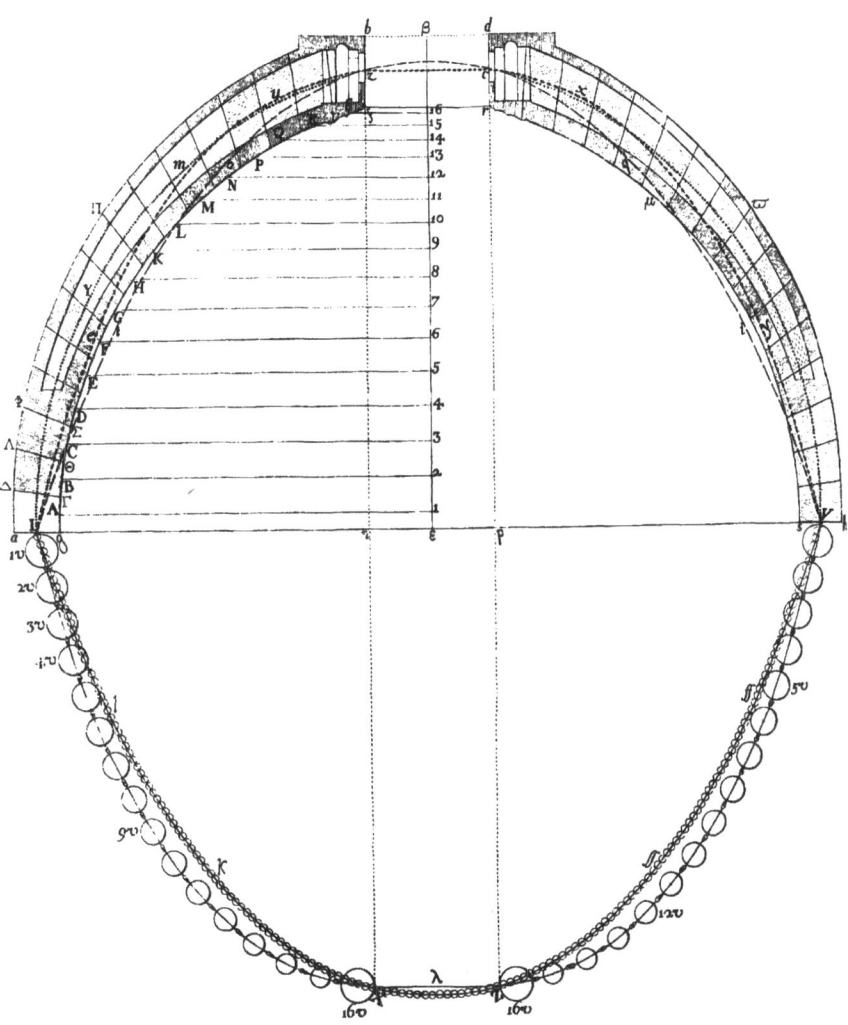

Figure 1. Poleni's test of the stability of the dome of Saint Peter's in Rome

It seems, reading Leibniz's answer to Jacob Bernoulli's question about the form of the catenary,

> The problem of the funicular or catenary, presents a double interest, first the one to enlarge the art of inventing, in other words Analysis, until now unable to affront correctly such questions, secondly the one to make the technique of building progress [Leibniz 1691: 243],

that the affinity between vault and catenary was well known at that moment, so we wonder why Benvenuto, after having told us that de la Hire links both problems in his *Traité*, but, as Johann Bernoulli wrote to Leibniz in October 1698, "sees something but what he has seen he himself does not understand",[2] tells us that "the first to establish the affinity between arches and catenaries was another scholar, an Englishman, David Gregory" [Benvenuto 1991, II: 327].

It is important to distinguish, here as well as in what follows, different problems. There is a mathematical problem, that of solving the problem of the catenary or of the vault using differential calculus. And there are mechanical problems:

1) What particular model of vault would stand best if it has the form of a catenary? This is a modelisation problem. The vault has to be made of infinitely thin blocks perfectly rigid and without friction between them.

Infinitely thin, I say. But is it possible then to split the mathematical from the mechanical problems? We'll try, as at the worst it could bring us is to learn us something about the relations of mathematics with mechanics.

2) What are the mechanical similarities and differences between the vault and the catenary?

There are different answers to that question, depending on the fundamental principle you use:

a) The similarity is that the stresses are tangential in both problems. The difference is that those stresses are tensions in the catenary and pressures in the vault.

---

[2] "La Hirius in suo Tractatus Mechanico, Propp. 123, 124 & 125, affinitatem inter fornices & catenarias subolvisse videtur ; nec tamen rem satis assequi potuit, nostro Calculo destitutus. Vidit aliquid ; quid autem viderit, ipse non intelligit." *Virorum celeberrimorum, G. G. Leibnitii et J. Bernoullii, Commercium philosophicum et mathematicum,* Lausanne and Geneva, 1745, vol. 1 letter 82, p. 412.

b) If one uses a variational principle to solve the problem, the similarity is that the equation to solve is the same, while the difference is that the catenary corresponds to the minimum and the vault to the maximum solution.

This said, let us go back to the historical problem. For Leibniz, the reason why de la Hire could not understand that affinity is that he was *lacking our calculus*. But couldn't he see the mechanical affinity just as Hooke could have seen it? De la Hire just says, "This proposition is only a converse of the previous one that was about a hanging rope." I would say that de la Hire saw a mechanical affinity but could not formulate it because, as we will see in the next section, he could not formulate completely the idealised model of either the catenary or the vault. But I don't agree with Leibniz that the reason why de la Hire couldn't express the affinity is due to his lack of knowledge of the calculus. The analogy is expressed by calculus, but its mechanics can be understand without it, even if infinitesimals are implied.

De la Hire saw that the important point is the direction of the pressure. He says, "As the courses of the stones or voussoirs are supposed to be infinitely polished, one does only have to consider them for the direction of the weight of the voussoirs."[3] In fact, it was first realised that if the pressure doesn't follow the curve of the vault, the stones would spring out of it, just as the parts of the chain in the catenary would do if the tensions didn't follow the curve. To follow the curve they have to be tangential. That way of thinking will become clearer reading Gregory.

The fact that Johann Bernoulli does not speak of the vault in his lectures to de l'Hôpital [Bernoulli 1741] and so neither does de l'Hôpital in his *analyse des infiniment petits*, also shows that the affinity is a mechanical one and doesn't bring anything to the calculus.

Let's turn now to Gregory's text on the catenary, written after he had read the three answers to Jacob Bernoulli's question of May 1690 by Johann Bernoulli, Leibniz, and Christiaan Huygens [Gregory 1697]. As Truesdell says, "For 1690, these three solutions, in the order received [Johann Bernoulli, Leibniz, Huygens], exhibit the mathematics of the future, the present, and the past" [Truesdell 1960: 66]. In fact Johann Bernoulli and Leibniz used differential calculus but just published the results, and Huygens didn't use it. To know how Bernoulli and Leibniz used the calculus we have to go to letters by Leibniz and to Bernoulli's lectures to l'Hôpital [Bernoulli 1741], texts that Gregory couldn't have read. Gregory's aim is to show that he could demonstrate the same results with the help of Newton's calculus of fluxions.

---

[3] "Puisque les lits des pierres ou voussoirs sont supposés infiniment polis, on les doit seulement considérer pour la direction de la pesanteur des voussoirs" [De la Hire 1695: 467].

Cum problema de figura Catenae (id est lineae flexilis, versus centrum longinquum gravis, & pondere suo dum duobus extremis immotis dependet incurvatae) sit inter hujus aevi Philosophos imprimis nobile, ac à Celeberrimis Viris Hugenio, Leibnitio & Bernoullio, plurimae figurae stius proprietates fuerint detextae, & in Actis Eruditorum Lipsiae (at sine demonstratione) editae : Libuit harum omnium demonstrationes pertexere, ope Methodi Neutoniae Geometris hodie familiaris, fluxiones è fluentium relatione data determinandi & vicissim. [Gregory 1697: 637]

The sixth corollary of proposition II tells us that,

the catenary in a vertical plane but inverted, allows to find but doesn't govern the figure of the thinnest arch or vault. It is to say that very little rigid and polished spheres arranged to form an inversed catenary, constitute an arch of which no part is being pushed inside or outside by the other parts; but leaves those infinitesimal points fixed, supported in virtue of her form [Gregory 1697: 640; my translation].

This idea is nicely illustrated by Poleni (Figure 2). The affinity of the vault and the catenary is common knowledge for Bélidor, in his *Science des ingénieurs* [1729], as one can see from the way he advises to design arches:

Si l'on veut construire une voûte naturelle dont la largeur et la hauteur soient données, il faut sur une surface verticale tracer une ligne CD égale à la largeur de la voûte, abaisser du milieu de cette ligne une perpendiculaire EF égale à la hauteur qu'on veut lui donner, ensuite attacher l'extrémité d'une chaîne au point C, et porter l'autre extrémité vers D, de manière qu'en augmentant ou diminuant la chaîne, son propre poids la fasse passer par le point F lorsqu'elle sera arrêtée aux endroits C et D. Après cela on pourra, avec un crayon que l'on conduira tout du long de la chaîne (sans pourtant vaciller), tracer une courbe, et là-dessus établir la figure du faux cintre de la voûte, la coupe des voussoirs et le reste [Bélidor 1729: 149-150] (Figure 3).

To understand the equilibrium of the catenary and of the vault, Hooke probably, and Gregory certainly, are considering the equilibrium of the forces. They use the fundamental principle:

$$\sum \vec{F} = 0,$$

or the parallelogram law. But we know that there is an other way of solving the problem, using the principle of virtual works. It is more difficult to tell when

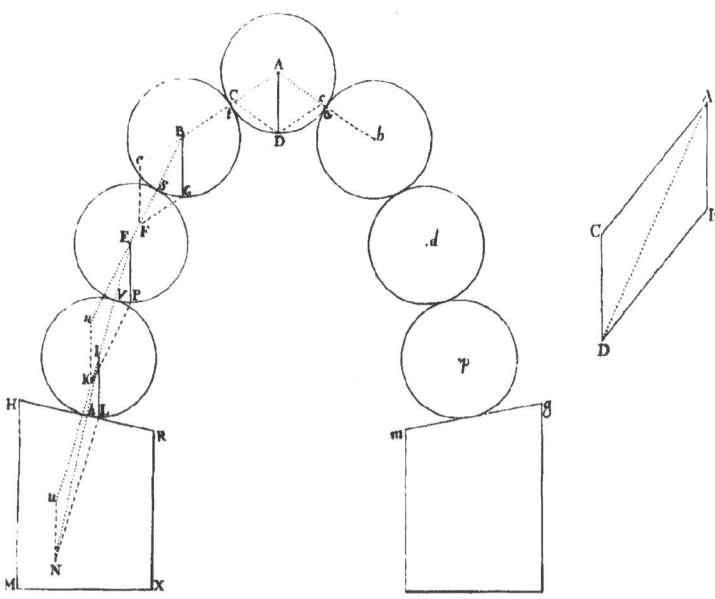

Figure 2. Poleni's illustration of the best form to give to a vault

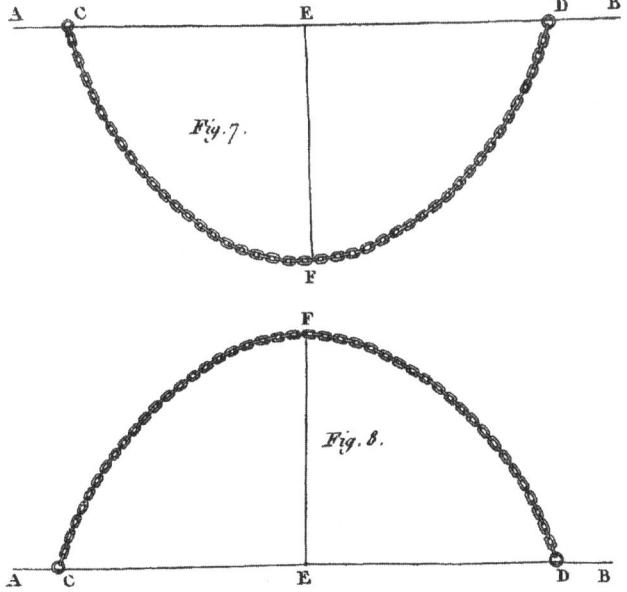

Figure 3. Bélidor's illustration of best form to give to a vault

people understood the similarity between the vault and the catenary in terms of energy.

In 1785, Lorenzo Mascheroni insists on that analogy, as one can see in his figures, but he doesn't explain it [Mascheroni 1785].

Finding the right model for the catenary and the vault

If one loads a rope or chain with various weights, one can obtain different forms. Pardies showed in 1673 that one could also obtain a parabola if, instead of a chain of uniform weight having *ds*, an element of the curve, of constant mass, we suspend to the chain weight such that *dx*, an element of the horizontal axis, is of constant weight, such as in the suspension bridge [Pardies 1673]. The same is true for the vault. One can make it circular and stable if one correctly chooses the weight of the different voussoirs. In fact, one has to precisely describe the model of the catenary or of the vault, that is, one has to select particular constitutive equations.

For the catenary

The hanging rope will take the form of a catenary, the hyperbolic cosine, if one takes a chain of uniform mass, perfectly flexible and of constant length. Pardies doesn't state that explicitly, nor does Johann Bernoulli in his public answer to Jacob's question but he does in his lectures to de l'Hôpital: "The thread, the rope, the chain or whatever represents the curve is supposed to be perfectly flexible in all its points and inextensible" [Johann Bernoulli 1741: 492].

For the vault

To get the same form, the vault ought to be infinitely thin, the voussoirs perfectly rigid and polished so that they slide one upon the other without friction.

The model gives a mathematical description of the constitution of the voussoirs: they are infinitely thin, the distance between two points is constant and we neglect friction.

De la Hire considers the vault as circular and the voussoirs as equal and reduced to their centre of gravity. Usually architects such as de la Hire, Couplet and Bélidor consider the catenary to be the curve linking the centre of gravity of a vault having a certain thickness. Mathematicians such as Jacob Bernoulli and David Gregory consider the vault as being infinitely thin.

Fundamental quantities

Both problems helped people to understand better some fundamental quantities for mechanics and for geometry from the most difficult ones.

Mechanical quantities: Tension and pressure [Truesdell 1968: 206]

Pardies's study of the catenary describes the tension with care and shows its vectorial character, even if he denominates it differently:

> *Force de leur traction: Ces corps suspendus étant inclinez, tireront diversement les cordes qui les soûtiennent; & la force de la traction se mesure comme dans l'article 67 en prenant dans la ligne de direction[4] au point F, & tirant les parallèles FG, Fg. Car la force du poids on étant exprimée par la ligne FB, la ligne BG exprimera la force dont la corde oA est tirée, & la ligne Bg celle de la corde na. Ce qui se peut exprimer encore par les deux poids D & d qui seroient au corps no, comme les lignes BG, Bg à la ligne BF* [Pardies 1673: 177].

Pardies explains clearly that the tensions in the chain must be tangential. This property was very important for Johann Bernoulli's solution of the catenary.

The affinity between both problems showed that tension and pressure were two manifestations of a more general notion, the stress.

Jacob Hermann generalized both notions in his *Phoronomia*:

> *The tension or compression of a thread or body at any of its points or at an element of the curve is that force of the thread or body which resists that power or force growing from all the applied powers and tending by pulling the thread in opposite directions to tear it apart. This tension exactly equals or is equipollent to that tearing force resulting from all the powers applied to the body* [Hermann 1716].

Geometrical: the ray of curvature

This important mathematical tool was known long before the period we are considering and Christiaan Huygens had already showed how useful it could be for mechanics. Jacob Bernoulli had discovered its differential expression working on the logarithmic spiral but published that expression with his study of the elastica, underlining in this way its importance for mechanics. As we will see

---

[4] The 'line of direction' was a technical term introduced by Tartaglia [1554: 84] to account for the vectorial character of the weight. It was generalized to other forces by Roberval in 1636.

later, that importance is shown once more by the completely original use he makes of the ray of curvature in the study of the vault [Radelet 1999a].

Fundamental principles

Purely geometrical

As Benvenuto shows clearly [Benvenuto 1991: 313-315], before de la Hire and his followers, vaults and even the greatest domes were built following purely geometrical rules such as the one given by Derand, still quoted by Bélidor in 1729 (Figure 4). Blondel's treatise on the *Résolution des quatre principaux problèmes d'architecture* [1673] is also purely geometrical, and so is a rather difficult paper by de la Hire, *Remarques sur la forme de quelques arcs dont on se sert dans l'Architecture* [1702].

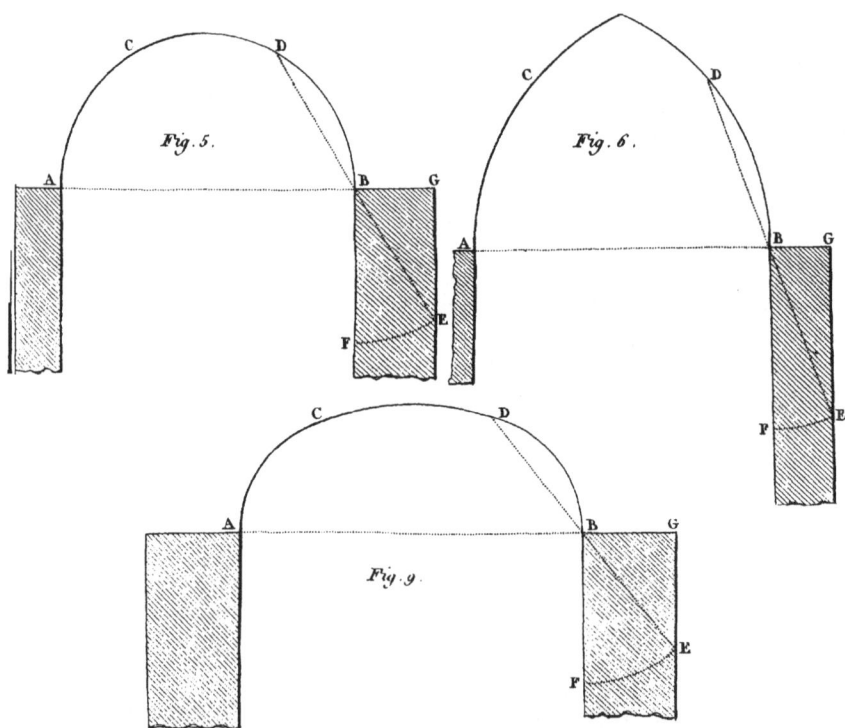

Figure 4. Derand's law to find the thickness of a pillar supporting an arch

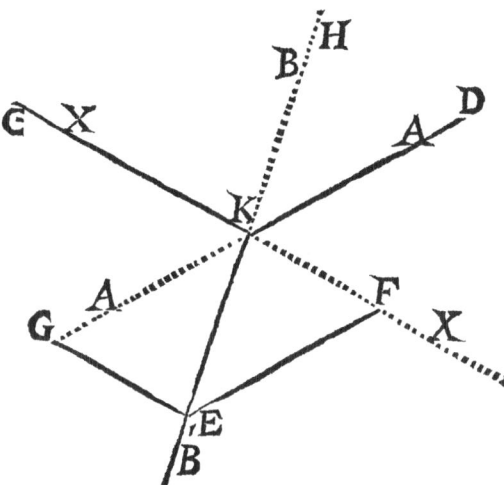

Figure 5. De la Hire's figure for proposition XXI

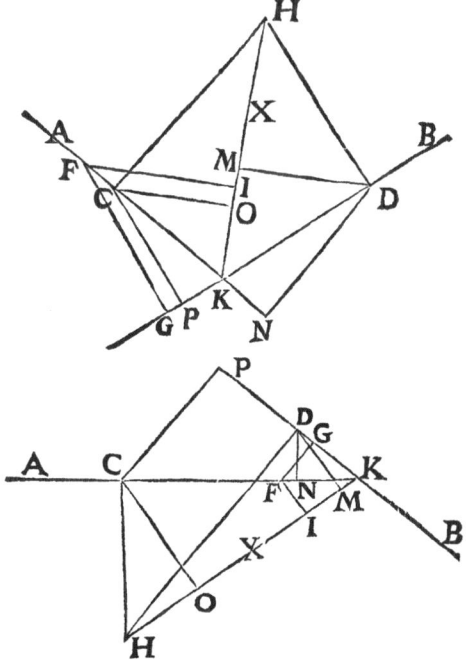

Figure 6. De la Hire's figure for proposition XIX

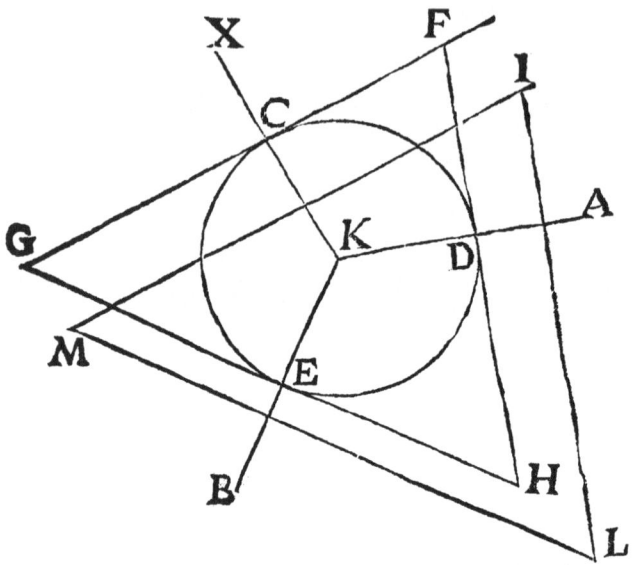

Figure 7. De la Hire's form of the parallelogram law

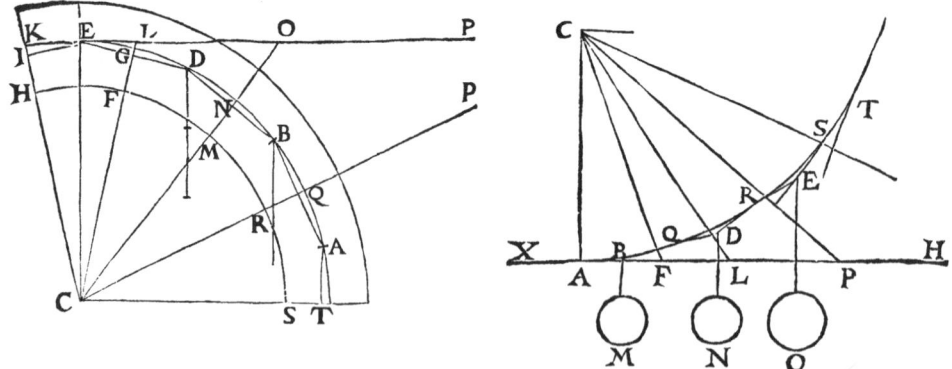

Figure 8. De la Hire's figures for the best vault and the catenary

The law of the lever or the sum of the moments of momentum must be zero

Truesdell [1968: 239-271], and more recently Giorgio Maltese [Maltese 2001], showed how difficult the way was towards the general law of moment of momentum. But a special case of that law was the first mathematical law of the whole of mechanics. It is called the law of the lever and up until the period we are considering it was the only mathematical law in mechanics. It is perhaps for this reason that the catenary couldn't be treated mathematically before the invention of calculus, which allows the use of the property that the tensions are tangential.

Nevertheless, I don't think anybody studied the problem of the catenary with the help of the lever. On the contrary, de la Hire's first attempt to treat the equilibrium of the vault is based on the law of the lever. In fact we must be more precise. De la Hire's mechanics rests upon one mechanical principle, "le levier qu'on peut regarder comme la proposition fondamentale de toute la Mécanique, puisque les autres parties s'y peuvent réduire facilement [De la Hire 1695 : Preface]. From the law of the lever, de la Hire derives a particular form of the parallelogram law of forces. His proposition XXI says, "Il faut trouver trois puissances AXB, qui tirant un point K par trois directions données CK, DK, EK, soient en équilibre entre elles" [De la Hire 1695: 70] (Figure 5). To do so, he traces the following figure where one might think he uses the parallelogram law, but instead he uses the XIXth proposition, which says,

> *dans un levier angulaire CHD, la puissance A sera à la puissance X comme le sinus de l'angle DKH fait par les directions des deux puissances X & B, ou de son supplément qui est le même, au sinus de l'angle CKD, ou de son supplément DKN qui est le même, fait de la direction des deux puissances A & B; & la puissance B sera à la puissance X, comme le sinus de l'angle CKH au sinus de l'angle CKD ou de son supplément CKP, ou CKV qui est le même sinus* [De la Hire 1695: 63-64] (Figure 7).

Then, de la Hire gives another form to the theorem equating the proportion of the forces to that of the sides of a triangle having its sides perpendicular to those forces (Figure 7). That last theorem is the one he uses to solve the problem of the vault (Figure 8) and to deduce that the weight of the different voussoirs have to be proportional respectively to KL, LO, OP on an horizontal line tangent to the top E of the vault.

Antoine Parent also studied vaults and catenary but, as Truesdell [1960: 80, 109-110] explains, his works are difficult to find. In fact one finds in *Histoire de l'Académie de Paris* a short remark without mathematics saying that Parent had found the curvature of the external part of a vault, the internal one being circular and the weights of the voussoirs different from each other and given by de la Hire's law [Parent 1704]. One should check with the *Essai* of Parent [1713].

Jacob Bernoulli also applies the law of the lever twice to solve that problem [Radelet 1994a]. First he uses it to compute the tangential pressure exerted by the top of the vault BEF on the voussoir KDEL, rather than using the parallelogram law. This is rather similar to what de la Hire did. He imagines that the top BEF of the vault is acting on the lever EF. Thus the weight of BEF, $2s$, is acting on M, the middle of the lever, that sits under the centre of gravity, E of BEF. This is equivalent to half the force acting at twice the distance, thus s is acting in E. But s does not act vertically, it acts along LI. So LI : s = FE : FI. But as EFI and EDN are similar we have LI : s = DE:DN = ds : dx. Thus he finds that the pressure along LI is $\dfrac{sds}{dx}$. Another force acts on the voussoir; to find that one, Jacob imagines a file maintaining the weight ds of the voussoir KDEL along the perpendicular to the vault OP. The tension in that file is to the weight ds of the voussoir as DN to DE or as dy to ds. So the tension is dy (Figure 9).

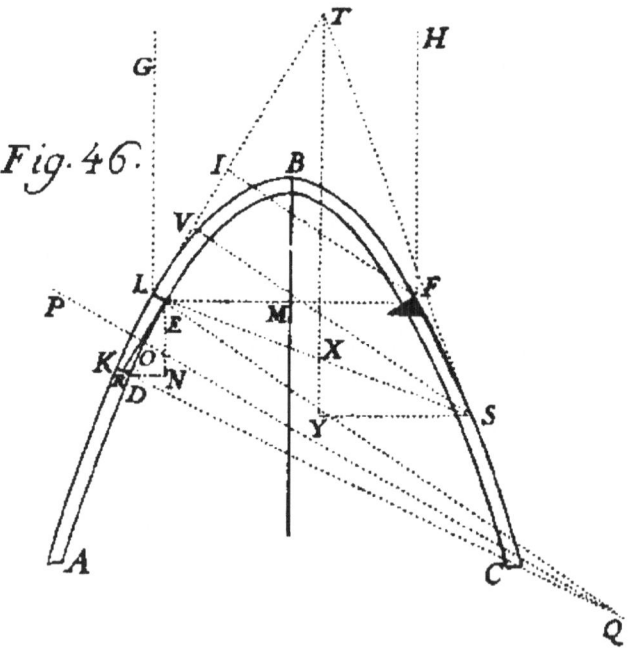

Figure 9. Jacob Bernoulli's figure for the study of the best vaults

But then he uses it in a completely different way. He imagines that the equilibrium of the vault could be broken if the voussoir KDEL rotates around the point D fixed by friction. To avoid that, he has to maintain in equilibrium the lever DE on which two forces are acting, $\dfrac{sds}{dx}$ and dy. The first one acts on E along IL or RD, and the other one on O, the middle of the lever along PO. If one applies the law of the lever, one gets

$$\frac{sds}{dx} \cdot DR = dy \cdot OD \,.$$

As Cramer points out, Jacob makes here an error saying that

$$DR = \frac{DE^2}{DQ} \ \ instead \ of \ \frac{DE^2}{2DQ} \,.$$

For that reason, Jacob doesn't find the catenary as a solution of that problem. Apart from that mistake, Jacob Bernoulli uses the balance of momentum correctly.

The parallelogram law of forces or sum of the applied forces is zero

The catenary. Stevin discovered the parallelogram law studying the equilibrium of a thread loaded with some weights as we can see in his figures dating from 1605 (Figure 10).

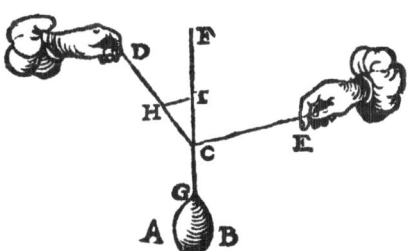

Figure 10. One of Stevin's first figures for the parallelogram law

The law was then given as a trick by Roberval at the end of his *Traité de Méchanique des poids soutenus par des puissances sur les plans inclinés à l'Horizon,*

> *Si, de quelque point pris en la ligne de direction du poids, on mène la ligne parallèle à l'une des cordes jusqu'à l'autre corde, les côtés du triangle ainsi formé seront homologues au poids et aux deux puissances* [Roberval 1636: 28].

It is only in 1687, with the *Principia* and in Varignon's *Projet d'une Nouvelle mécanique ou statique* that that law came into its real importance. In fact Varignon observes that one had always based statics, that is to say, the simple machines, upon the law of the lever. He proposes to base it on the parallelogram law. We know that in fact one needs both laws to do statics.

After Varignon's publication, one sees Jacob Bernoulli struggling in his *Meditationes*, trying to understand it and confronting the results it gave with the one obtained with the law of the lever [Radelet 1989]. But we saw in his treatment of the vault that even at the end of his life, he still preferred the law of the lever, even to find the component of a force.

In another of his last *Meditationes*, Jacob tries to generalise the problem of the catenary studying the equilibrium of a rope under the action of various forces acting in different directions:

> *Filum ACDEFGB extremitatibus suis A & B suspensum ab infinitiis potentiis C, D, E, F, G, juxta directiones quasvis HC, HD, IE, KF, LG agentibus extenditur. Quaeritur fili curvatura, ejus directio media LP, & vis, qua secundum LP impellitur* [Jacob Bernoulli 1744b].

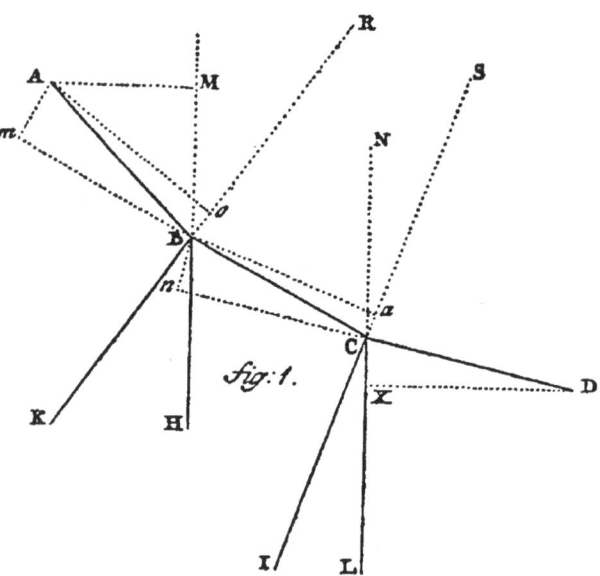

Figure 11. Daniel Bernoulli's figure for the study of the catenary

This argument will be taken up again by Jacob Hermann in his *Phoronomia* [Hermann 1716: bk. I, sec. I, chap. iii] and in 1728 by both his nephew, Daniel Bernoulli [1728], and by Euler [1728]. But there things are quite different. Two years before, Daniel had given a demonstration of the parallelogram of forces [1726] and he will use it to find the components of the different forces along a first direction orthogonal to the curve and an other one always parallel (vertical) to a fixed direction. So he easily finds the particular cases of the catenary if only the components parallel to each other are different from zero or the *lintearia* if only the components perpendicular to the curve are different from zero (Figure 11).

After having read Daniel's article, Euler, in the same volume of the *Commentarii Academiae Petropolitanae*, gives a program for the solution of the problem of finding the curve assumed by an arbitrarily elastic band loaded by arbitrary forces at its several points. But it will take him fifty years to fulfil it completely. Truesdell says about Daniel that his article contains nothing new:

> *That he was ignorant of his uncle's unpublished work is to be expected, since Jacob Bernoulli's paper were kept from Johann Bernoulli and his circle; that Daniel Bernoulli should not know the general solution in the book by his senior colleague, Hermann, is surprising, especially since Euler refers to it* [Truesdell 1960: 146-147].

But in my opinion, Daniel added something to what Hermann had done. He choose two particular directions on which he tacks the components of the various forces, so he doesn't have the difficulty Hermann had. Truesdell explains, "the difficulty may lie in the failure to realize that his polar co-ordinate diagram must be drawn over again at each point" [Truesdell 1960: 86]. So Daniel's choice of two particular directions, a choice he made in order to find the well-known particular cases easily, will open the way of generalisation for Euler. He'll change one thing; he will replace the components parallel to a fixed direction by tangents to the curve, obtaining an orthogonal system of reference.

The vault. We saw that de la Hire demonstrated his particular form of the parallelogram law from the law of the lever. Couplet on the contrary starts with the parallelogram law. But it was sufficiently new to be mentioned in the introduction to Couplet's article written in the *Histoire de l'Académie de Paris*:

> *Son impulsion [de la clef de voûte] sur ce voussoir ne peut être qu'une perpendiculaire tirée du centre de gravité de la Clé sur la surface du voussoir. Cette ligne est en même temps la diagonale d'un parallélogramme dont les deux côtés seroient la tendance*

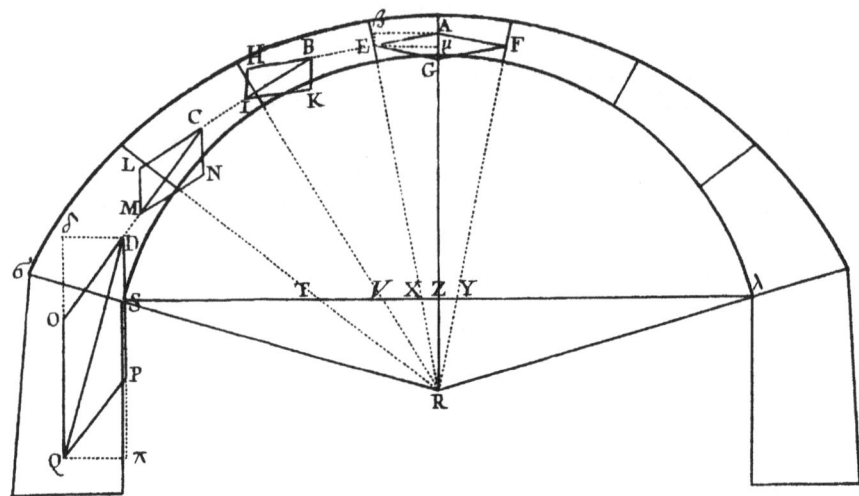

Figure 12. Couplet's figure for the study of the stability of vaults

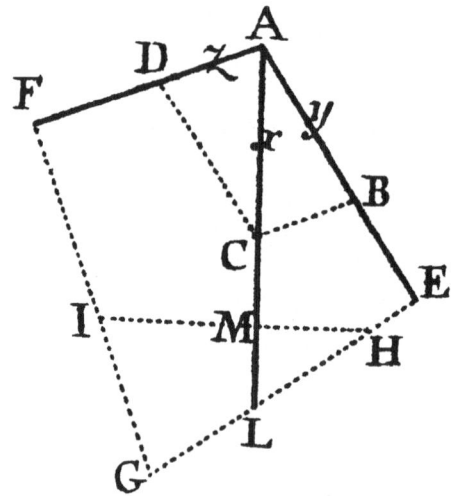

Figure 13. Couplet's form of the parallelogram law

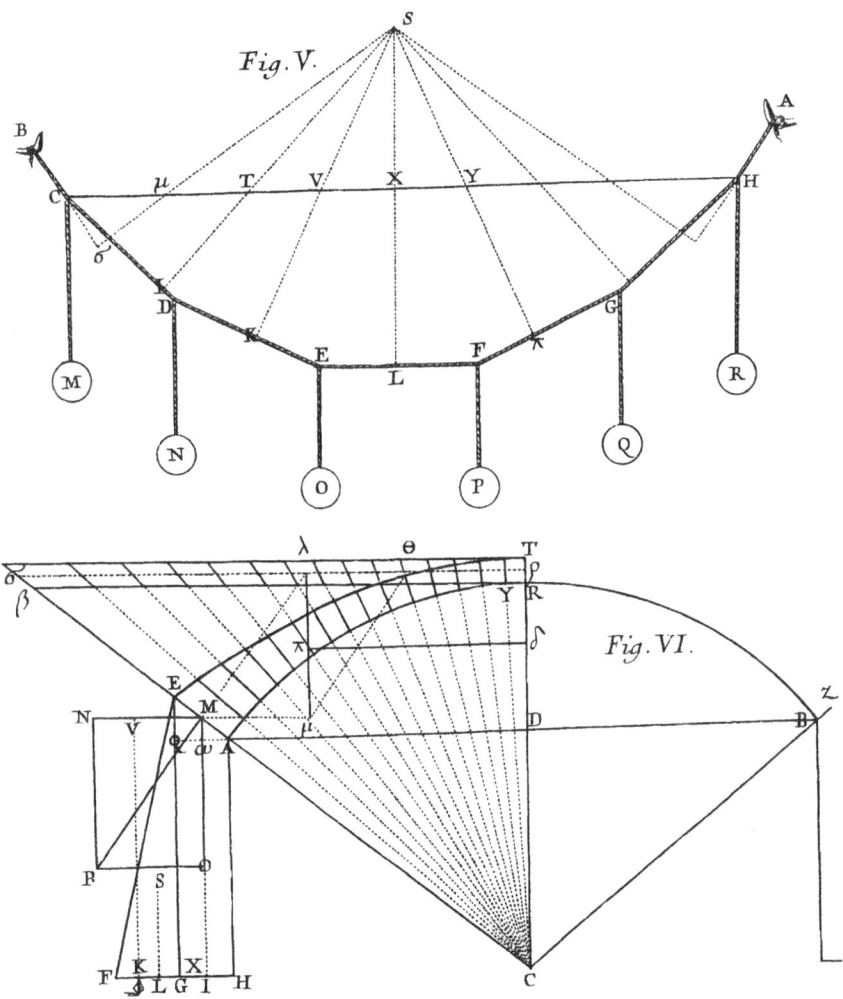

Figure 14. Couplet's use of de la Hire's graphical method

verticale de la Clé pour tomber, & un effort horizontal pour pousser le voussoir ou l'écarter [Couplet 1729: 76] (Figure 12).

But Couplet prefers de la Hire's form of the parallelogram law:

Si la force x se décompose en deux forces y & z, ces trois forces seront entre elles comme les côtés d'un Triangle formé par les perpendiculaires menées sur les directions de ces trois forces [Couplet 1729: 80] (Figure 13).

The reason is that it lead to a very easy graphical law to find the weight of the different voussoirs of a stable vault:

Lorsque tous les joints de la Voûte sont dirigés vers un même point R, comme l'on dit qu'ils doivent être dans la Voûte, dont l'intrados est circulaire, si l'on tire une ligne horizontale quelconque ST, les pesanteurs des Voussoirs A, B, C, D, seront exprimées par les parties XY, VX, TV, ST, de cette ligne horizontale renfermées entre les joints prolongés de ces Voussoirs [Couplet 1729: 86].

In Figure 14, we can see that Couplet uses de la Hire's graphical method and that he is trying to consider infinitely small voussoirs, in order to get a continuous curve. He uses differential calculus in his work but couldn't find the equation of the curve of the extrados. To the question, "Déterminer la courbure uniforme d'une voûte, telle qu'elle se maintienne en équilibre, & dont nous considérons les voussoirs comme polis, c'est-à-dire, sans liaison [Couplet 1729: 95], he answers, "Prenés une corde garnies de poids égaux…" One explanation could be that he doesn't seem to have the notion of ray of curvature and so always uses circles.

The same year, Bélidor, in his *Science des ingénieurs* [1729] dedicates an entire book to the mechanic of vaults. After a survey of ancient geometrical methods, the book starts with Principles taken from the mechanics. The first one is the parallelogram law and de la Hire's variant (Figure 15). And as we have already said, to find the best curve for equal voussoirs he takes the chain.

Bouguer and Bossut are familiar with both the parallelogram law and the ray of curvature as one immediately sees in their illustrations (Figures 16-17).

The necessity of considering both

Truesdell often insists on the fact that one has to consider both the $\sum \vec{F} = 0$ and $\sum \vec{M} = 0$ and from that point of view, I think that he doesn't insist enough on Jacob's way of reasoning for the most stable vault. We have already seen that

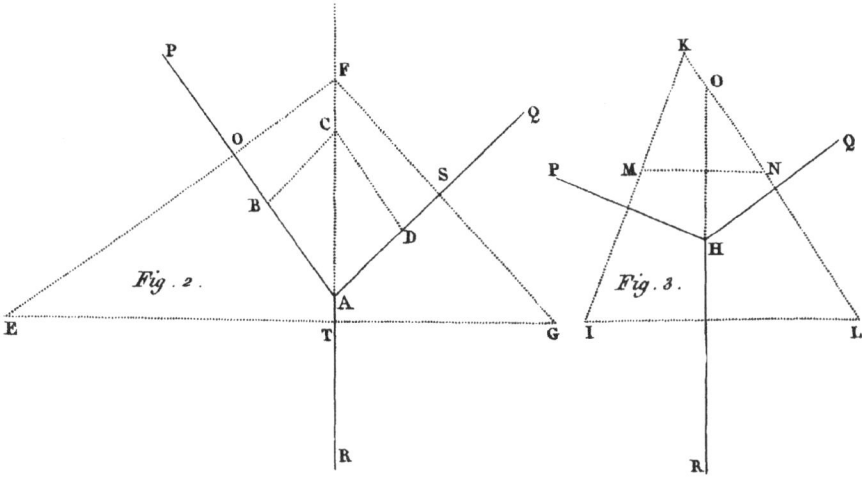

Figure 15. Bélidor's figures for the parallelogram law

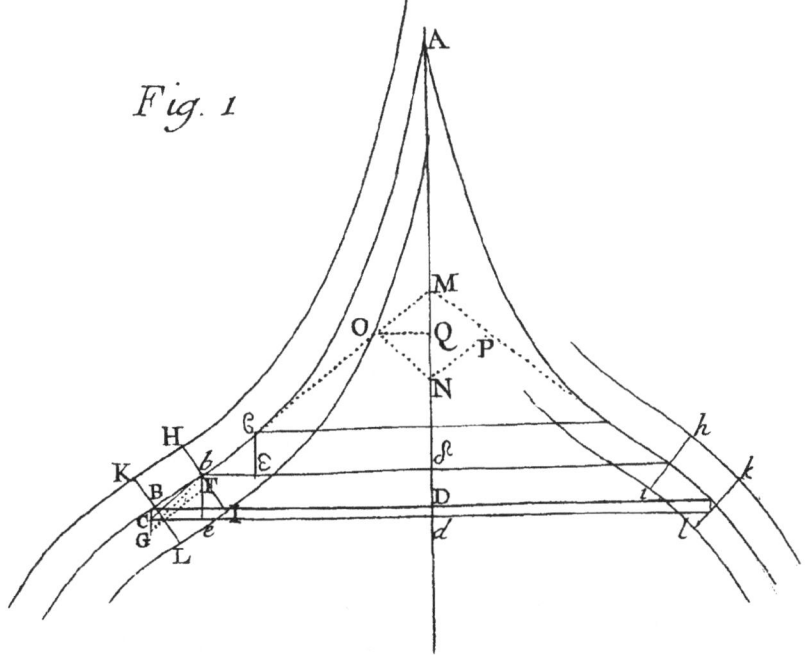

Figure 16. Bouguer's figure for the study of the equilibrium of a vault

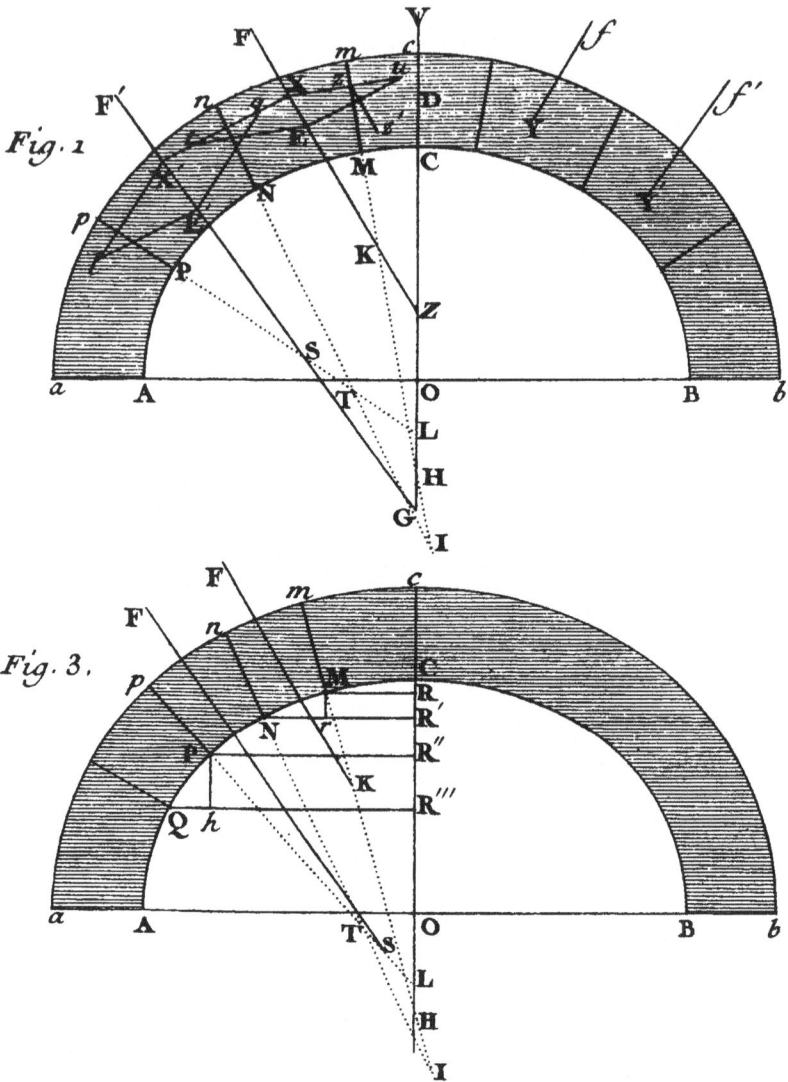

Figure 17. Bossut's figures for the study of the equilibrium of a vault

he used the balance of momentum but let's go back to the beginning of his work where he says:

> *Je suppose que si l'on retirait la partie EBF de la voûte, la pierre DL n'adhérant à la partie inférieure AK par aucun ciment, ne peut tomber que sous l'effet de son propre poids; ce qu'elle ne peut faire qu'en glissant sur le plan KD, si ce plan est parfaitement lisse, ou en tournant autour du point D, ... si il y a friction avec DK* [Jacob Bernoulli 1744c].

In other words, the voussoir can only have two movements, a movement of translation, sliding on the surface KD, or a movement of rotation around D. To prevent sliding Jacob uses the balance of forces, a law that the engineers call 'equilibrium to translation', and to prevent it to rotate, he uses balance of momentum, 'equilibrium to rotation', the engineers say. (It would be interesting to find out who introduced those two expressions, equilibrium to translation and to rotation, into the language of engineers but I think it's rather recent.)

Minimal energy

When the young Huygens started to study the problem of the catenary in 1646 following a suggestion made by Mersenne, he immediately gave an axiomatic form to his study. And the first axiom he used, after Torricelli, is an axiom of minimal energy. Torricelli wrote in 1644, "Nous poserons en principe que deux graves, liés ensemble, ne peuvent se mouvoir d'eux mêmes, à moins que leur commun centre de gravité ne descende [Torricelli 1644: bk. 1, prop. I, 99 ; translated in Duhem 1906, 2: 3]. Huygens changes that dynamic in a static principle: The centre of gravity occupies the lowest possible place. With the help of this axiom he demonstrates that the centre of gravity of two weight lies on a vertical passing through the intersection of the prolongation of the adjacent sides, and with the help of that property traces the curve point by point. (Figures 18-19).

Johann Bernoulli once used this hypotheses in his lectures to de l'Hôpital for demonstrating the last property 13,

> *Si l'on se représente une infinité de courbes tracées sur EF, égales à la funiculaire EBF, que l'on étend en une droite, et que l'on met en ordonnée en chacun des points de chaque [courbe] étendue des droites égales à leurs distances respectives de la ligne EF, parmi toutes les aires qui sont ainsi formées, celle qui est engendrée par la funiculaire sera la plus grande.*

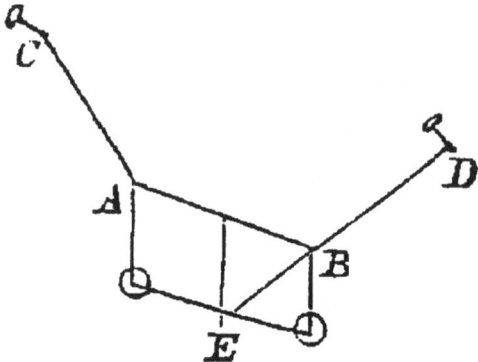

Figure 18. Huygens's first figure for the study of the catenary

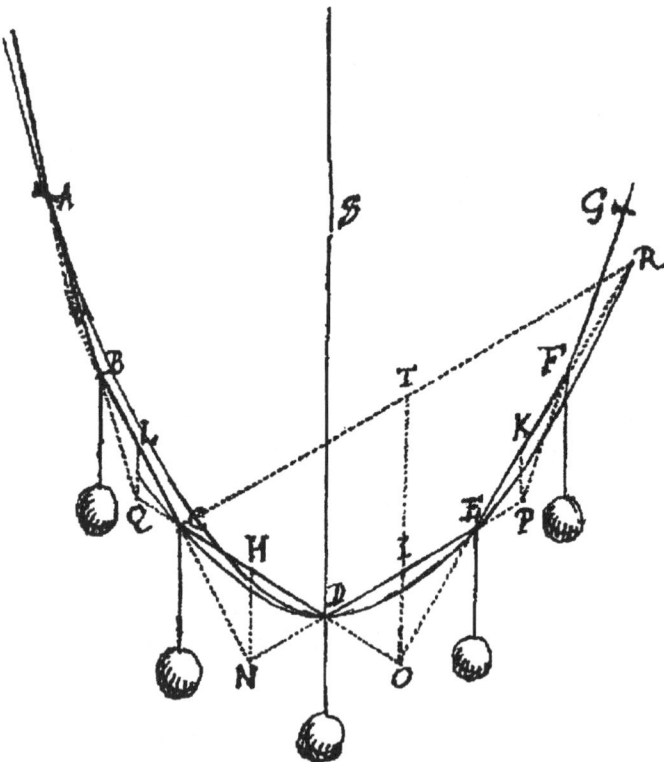

Figure 19. Huygens's second figure for the study of the catenary

Ce que l'on montre au moyen de cet axiome que le centre de gravité occupe l'emplacement le plus bas possible [Johann Bernoulli 1741: 497],

but he doesn't give the demonstration.

Jacob is much more interested in this axiom and in all other extremal properties he encounters. Already in his first text on the Elastica, he enunciates without demonstrating them the following properties:

parmi toutes les isopérimètres qui s'appuient sur A f l'élastica est celle dont le centre de gravité est le plus éloigné de Af alors que la caténaire est celle dont le centre de gravité est le plus éloigné de la courbe elle-même [Jacob Bernoulli 1694: 599].

In his *Meditationes* CCXXXIX, *Inter omnes Figuras Isoperimetras vel aequalium arearum invenire illam, quae habeat maximum vel minimum aliquod Methodus Nova,* written in 1697, one finds various interesting examples:

1) Beyond all isoperimetres, find the one of which the centre of gravity occupies the lowest place. He finds the catenary.

2) Beyond all isoperimetres, find the one which contains the largest surface. He finds the circle.

3) Beyond all isoperimetres, find the one of which the centre of gravity of the contained surface occupies the lowest place. He doesn't find the elastica as he thought, but the catenary.

4) Beyond all curves containing the same surface, find the one of which the centre of gravity of the surface occupies the lowest place. He finds an impossibility.

Truesdell told us of the influence of those ideas on Euler and how with the help of Daniel Bernoulli, he followed that way to solve the problem of the elastic ribbon in his *additamentum* to his *Methodus inveniendi lineas curvas maximi minimive proprietate gaudentes, sive solutio problematis isoperimetrici latissimo sensu accepti* exactly at the same time as Maupertuis published his first article on his principle of least action.

Virtual works

The catenary

I showed in some details, in two articles how Huygens progressively and principally studying the catenary begins to use the principle of virtual works.

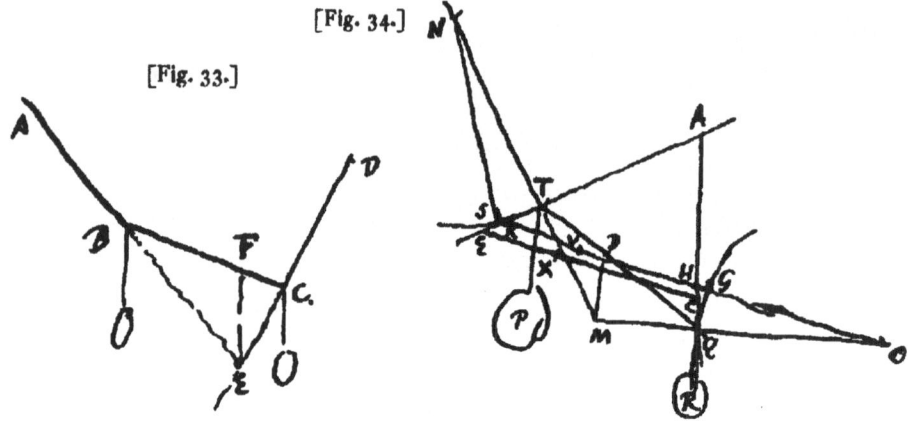

Figure 20. Huygens's figures for the study of the catenary using virtual works

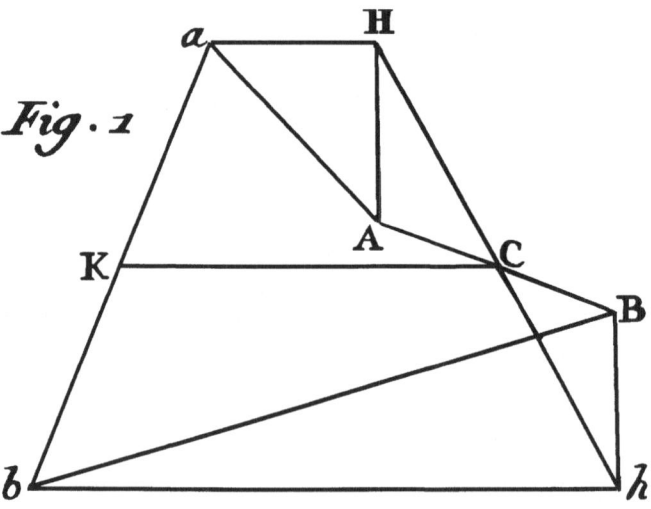

Figure 21. Mascheroni's figure for the elaboration of a principle of virtual works

The vault

As regards the vault, let return once more to Jacob Bernoulli's text. Truesdell says that for preventing the voussoir from sliding, Jacob uses virtual works. He first translates Bernoulli:

> The [infinitesimal] stone KL in figure 29 ... is to be regarded as a wedge trying to force itself into the triangle DQE. As it comes from KL into the position DE, that is, while it traverses the space KD, it pushes back the force pressing along IL by a distance KL-DE [Truesdell 1960: 83].

The idea of comparing the voussoir to a wedge was used the same year by Parent but Bernoulli couldn't have known that. Truesdell continues:

> Then the virtual work done by the normal force $-F_n$ pointing inward equals that done against the compression $-T$. That is, $-F_n$. $KD = - T.(KL-DE)$ [Truesdell 1960: 83].

This is, with other notations, exactly the equation written by Bernoulli *by the nature of the wedge*.

The champion of virtual works within the frame of the study of stability of vaults arrives much later: in 1785, Lorenzo Mascheroni wrote his work on the equilibrium of the vault. The work is based on the following theorem:

> If the weight B goes to b on an plane inclined on the horizon by an infinitesimal space Bb and during the same time by the way of a machine, the weight A goes to a along the inclined plane Aa, One traces the horizontals aH, bh and the perpendiculars AH, Bh; the force of the weight A will be to the force of the weight B in the composed reason of the weight A to the weight B and of the perpendicular travel AH to the perpendicular travail Bh [Mascheroni 1785] (Figure 21).

But even that one only considers the work done by gravity and vertical paths, just as Jordanus de Nemore already did for the inclined plane. And his idea comes directly from the Aristotelian conception of the gravity.

References

BELIDOR, B. FOREST DE. 1729. *Les Science des ingénieurs dans la conduite des travaux de fortification et d'architecture civile ... par M. Belidor*. Paris.

BENVENUTO, EDOARDO. 1981. *La scienza delle costruzioni e il suo sviluppo storico*. Florence: Sansoni. Engl. Trans. 1991. *An introduction to the history of structural mechanics. Part I: Statics and resistance of solids. Part II: vaulted structures and elastic systems.* 2 vols. New York: Springer.

BERNOULLI, DANIEL. 1728. Methodus universalis determinandae curvaturae fili. Pp. 62-69 in vol. 3 (1732) of *Academiae Imperialis Scientiarum Petropolitanae.*

BERNOULLI, JACOB. 1744a. Op. LVIII, Curvatura Laminae Elasticae. Ejus Identitas cum Curvatura Lintei a pondere inclusi fluidi expansi. Radii Circulorum Osculantium in terminis simplicissimis exhibiti, una cum novis uibusdam Theorematis huc pertinentibus, &c. Pp. 576-600 in *Opera Omnia,* Genève.

————. 1744b. Varia Posthuma N°. XI, Pp. 1036-1048 in *Opera Omnia,* Genève.

————. 1744c. Varia Posthuma N°. XXIX, Pp. 1119-1123 in *Opera Omnia,* Genève.

BERNOULLI, JOHANN. 1741. Op. CXLIX – Leçons à de l'Hôpital no. 36-45. Pp. 491-516 in *Opera Omnia,* vol. IV. Lausanne and Genève.

BLONDEL, F. 1673. Résolution des quatre principaux problèmes d'architecture. Pp. 355-530 in *Mémoire de l'Acad. Roy. Des Sciences* (1729), Paris.

COUPLET, C.A. 1729. *De la poussée des voûtes.* Pp. 79-117 in *Mémoires de l'Acad. Roy. Des Sciences* (1731), Paris.

DE LA HIRE, PHILIPPE. 1695. Traité de Mechanique in *Opera diverses ...,* Pp. 1-333 in *Histoire de l'Acad. Roy. Des Sciences,* vol. IX (1730), Paris.

————. 1702. Remarques sur la forme de quelques arcs don't on se set dans l'Architecture. Pp. 100-103 in *Mémoires de l'Acad. Roy. des Sciences* (1720), Paris.

DUHEM, P. 1906. *Les origins de la statique.* 2 vols. Paris.

EULER, LEONHARD. 1728. E 8, Solutio problematic de invenienda curva…. Pp. 64-77 in vol. 3 of *Academiae Imperialis Scientiarum Petropolitanae.*

GREGORY, DAVID. 1697. Catenaria. Pp. 637-652 in *Philosophical Transaction,* vol. 19, 1698. Repr. pp. 305-321 in *Acta Eruditorum Leipzig,* 1698.

HERMANN, JACOB. 1716. *Phoronomia.* Amsterdam.

HOOKE, ROBERT. 1675. *A description of helioscopes, and some other instruments.* London.

LEIBNIZ, G.W. 1691. De solutionibus problematic catenarii vel funicularis in actis junii A. 1691, aliisque a Dn. J.B. propositis. Pp. 435-439 in *Acta Eruditorum Leipzig.*

MALTESE, GIULIO. 2001. *Da "F=ma" alle leggi cardinali del moto: sviluppo della tradizione newtoniana nella meccanica del '700.* Hoepli: Milan.

MASCHERONI, LORENZO. 1785. *Nuove ricerche sull'equilibrio delle volte.* Bergamo.

PARDIES, G. 1673. *La statique ou la science des forces mouvantes.* Paris.

PARENT, ANTOINE. 1704. Sur la figure de l'extrados d'une voute circulaire, don't tous les voissoirs sont en equilibre entre eux. Pp. 93-96 in *Histoire de l'Acad. Roy. des Sciences* (1712), Paris.

————. 1713. *Essais et Recherches de Mathématique et de Physique*. Paris.

RADELET, PATRICIA. 2000. Déplacement, vitesses et travaux virtuels. Paper presented at the symposium "Varier pour mieux trouver", 4-9 September 2000, Patzcuaro. In preparation.

————. 1999a. La mesure de la courbure et la pratique du calcul différentiel du second ordre. Proceedings of the workshop, "Le pensée numérique", 7-10 Spetember 1999, Peyresq. In preparation.

————. 1999b. Sur les lignes courbes qui sont propres à former les voûtes en dome de Pierre Bouguer, 1734. *Sciences et Techniques en perspective*, 1st series, vol. 3, no. 2: 397-421.

————. 1994a. Le "de Curvatura fornicis" de Jacob Bernoulli ou l'introduction des infiniment perits dans le calcul des voûtes. Pp. 141-164 in *Between Mechanics and Architecture*, Patricia Radelet-de Grave and Edoardo Benvenuto, eds. Basel: Birkhäuser.

————. 1994b. Etude de l'Essai sur une applicatione des règles de Maximis et Minimis à quelques Problèmes de statique, relatifs à l'Architecture, par M. Coulomb. *Science et technique en perspective*, no. 27, University of Nantes.

————. 1989. La composition des mouvements, des vitesses et des forces. Proceedings of the Symposium "Der Ausbau des Calculus durch Leibnez und die Brüder Bernoulli", 15-17 June 1987, Bâle, in *Studia Leibnitiana* 17: 25-47.

ROBERVAL, GILLE PERSONNE DE. 1636. Traité de mechanique des poids sousternus par des puissances sur les plans inclinez à l'horizon. Pp. 1-36 in *L'Harmonie universelle* by M. Mersenne.

Sur les voûtes. 1729. Pp. 75-81 in *Histoire de l'Acad. Roy. Des Sciences*, 1731. Paris.

TORRICELLI, EVANGELISTA. 1644. *Opera Geometrica, De motu gravium naturaliter descendentium et projectorum libri duo*. Florence.

TRUESDELL, CLIFFORD A. 1960. *The rational mechanics of Flexible or Elastic Bodies 1638-1788*. Introduction to *Leonhardi Euleri Opera Omnia*, second series, vol. 11 (2). Zürich: Orell Füssli.

————. 1968. *Essays in the History of Mechanics*. Berlin and New York: Springer-Verlag.

Virorum celeberrimorum, G. G. Leibnitii et J. Bernoullii, Commercium Philosophicum et mathematicum. 1745. Lausanne and Geneva.

Rose Windows

Jacques Heyman[1]

The significant load on a large glazed window arises from the pressure or suction of wind. The masonry in a rose window resists this loading by developing flat-arch action, and thrusts with considerable force against the surrounding walls. Thus connexions at the periphery must be firm, and the masonry of the window itself must be of a satisfactory geometrical form to accommodate any irregularities which may develop at the boundary.

Introduction

It seems clear that a hole may be punched through a masonry wall with relative impunity. If the hole is rectangular, then the head will need a monolithic lintel to 'relieve' the weight of the masonry in the wall above; more sophisticatedly, the monolith may be replaced by a flat arch of several voussoirs. As the rectangular window becomes larger, then the flat arch develops, in the Gothic, into a pointed form, and the eye is given a good indication of the 'flow' of forces round the window.

If the hole is circular, then the opening has been called, for a century and a half, an *oculus*, or eye. The circular hole again needs reinforcement round the perimeter; the oculus is bounded by a circular arch in the upper half, and a reflected arch in the lower. Unglazed oculi admitted both light and air to large public rooms, whether in churches or private houses; they were not used in small chambers such as bedrooms where light was not needed and air could be admitted through an open doorway. Oculi can be found in the gables of Romanesque churches up to the late twelfth century, but larger round windows were already developing in early Gothic. An open oculus, of say 2 m diameter, admits some light, but not enough – it also admits rain.

As oculi became larger during the late Romanesque, so the need for primitive glazing arose, and this glazing required a frame. The simplest possible frame is formed by a single vertical member (a mullion), or by a mullion and a horizontal member (a transom), placed diametrally to form a cross. Further 'spokes' were soon added to form the wheel window, with 6, 8, 12, 16 or even more radial members running from a central 'hub' (which could be a glazed ring of stone) to the circumference [Cowen, 1979].

The static vertical loading on a spoke is negligible, whether the member runs horizontally, vertically, or is inclined. The critical loading on the members of a masonry window arises from transverse wind forces, and this loading is many

[1] 3 Banhams Close, Cambridge CB4 1HX, United Kingdom

times greater than the self-weight of the material. Thus gravity forces may be neglected for a primary analysis, and only the response to horizontal wind investigated.

The flat arch

A large window in a cathedral may be approximated by a rectangle, say 12 m x 9 m. There may perhaps be two main mullions, together with subsidiary mullions and horizontal transoms, but essentially the two main mullions receive the wind forces from the glazing housed into them, and transmit those forces to their ends at top and bottom of the window (Figure 1). The mullions may be thought to act as beams standing vertically and subject to a uniform horizontal load, just as architraves in Greek trabeated architecture may be thought to act as horizontal beams subject to their own vertical weight.

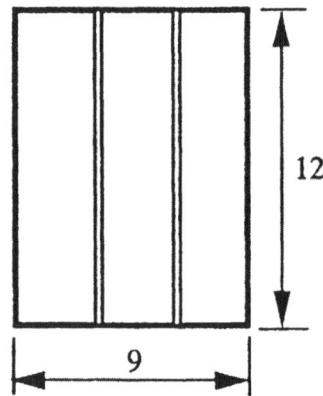

Figure 1. A rectangular window

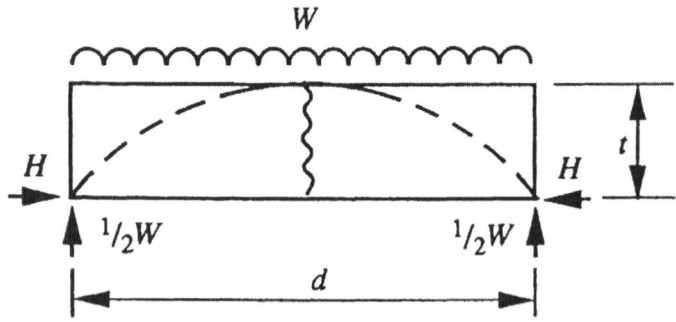

Figure 2. A Greek architrave

A simple architrave is represented in the sketch of Figure 2, and the (self-weight) load W induces vertical end reactions of ½ W. If the architrave is really acting as a beam in bending, then tensile strains will be present on the under surface, and corresponding compressive strains on the top. Such actions are in fact inadmissible for masonry [Heyman, 1995]; individual stones may be strong in tension but, once cracked, they can transmit only compressive forces between those portions which remain in contact. Many massive Greek architraves are cracked through near their centres, as indicated in Figure 2, and there is no possibility of regarding this cracked structure as a beam in bending – it is a flat arch formed of two voussoirs [Heyman, 1972]. Such a flat arch will remain stable if it is prevented from spreading; the horizontal thrusts H in Figure 2 are supposed to be provided (in a Greek temple) by adjacent architraves. A very small increase in span will induce a small sag in the architrave, and the central crack will develop into a 'hinge'; forces within the masonry, the thrust line, are indicated by the broken line in Figure 2.

Thus an uncracked architrave subjected to a uniform load W may be supported at its end by purely vertical forces ½ W. Once cracked, however, the architrave can only remain in equilibrium if a horizontal force H is developed; by simple statics,

$$H = \frac{Wd}{8t} \qquad (1)$$

Each of the two mullions in Figure 1 will be made from several pieces of stone, pinned together at their centres by metal dowels (of iron, phosphor bronze, stainless steel). The dowels can contribute little to bending strength, and the mortar (if any) in the joints can develop at best feeble tensions. The mullion is acting as a (vertical) flat arch composed of several voussoirs, the voussoirs being connected to each other and to the masonry frames at top and bottom by dowels, so that the stones cannot slip; the mullion is acted upon by (horizontal) wind forces, and behaves exactly as the Greek architrave of Figure 2. If the transverse wind pressure is taken as 1 kN/m² (20 lb/ft², a slightly low figure), then each mullion in Figure 1 will be subjected to a uniform loading of 36 kN (i.e., 3.6 tonnes). Thus if d/t is 24 (say), then $H = 3W$ from equation (1); a transverse load of 36 kN will engender end thrusts of 108 kN. Each mullion is sustained in equilibrium by horizontal reactions of 18 kN (the end reactions of ½ W of Figure 2), and the end dowels serve to pass these reactions to the main fabric; and the wind also induces vertical end reactions (H of Figure 2) of value 108 kN (if the mullion has proportions d/t = 24), as indicated in Figure 3. Each mullion will contain within its thickness a curved line of thrust (the broken line of Figure 2); this thrust line is quasi two-dimensional, lying in a vertical plane parallel to the direction of wind.

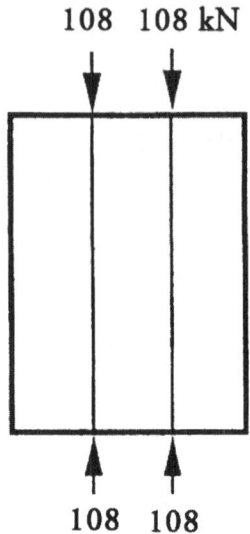

Figure 3. End loads on mullions

An obvious requirement is therefore that a mullion, and in general any member or sub-member of a framed masonry window, be essentially straight; that is, only gentle curves are permitted. The requirement is familiar; for example, the scissor braces connecting the crossing piers at Wells Cathedral are certainly not straight, but they are effective because straight 'props' may be contained within their profiles.

Equilibrium analysis

The above numerical example used, without comment, an assumption of a kind usually made by structural analysts. The wind acts upon the glazing of the window, and the glazing bars transmit the wind forces to the mullions, and so to the frame of the window; it was assumed that the loading on a mullion was distributed uniformly. Such as assumption is made in the absence of any specific information about the disposition of the internal framing of the window – in fact the problem is hyperstatic in the extreme, and the best the structural analyst can do in such cases is to make a 'reasonable' guess about some of the parameters entering into the calculations. Such guesses do not invalidate the process, and it is a matter of fact that structural engineers have always worked in this way, and have relied, unknowingly for the most part, on the comfort of the plastic theorems.

The lower bound, or 'safe' theorem, is simple – if the analyst can find a state of the structure in which the structure is 'comfortable', then there cannot exist any state which might lead to collapse. The power of this theorem is that the analyst is not required to determine the 'actual' state of a structure under the given loading – if any one equilibrium state can be found, then this is sufficient proof that the structure itself will exist in a (certainly different) state of safety.

Thus for the rectangular window of Figure 3 acted upon by a transverse wind load of 1 kN/m², each of the two mullions *could* be subjected to *uniform* loading of 36 kN, and this loading will engender the end thrusts shown of 108 kN. The forces shown in Figure 3 represent one possible state of equilibrium for the window – if the end thrusts can be sustained by the mullions (and this is a question of stress analysis rather than overall structural behaviour), then the calculations have led to the assurance that the window can carry the required wind load. (For a mullion of suitable dimensions in Figure 3, the average compressive stress arising from the end thrust is about 1 N/mm², which is an acceptable value.)

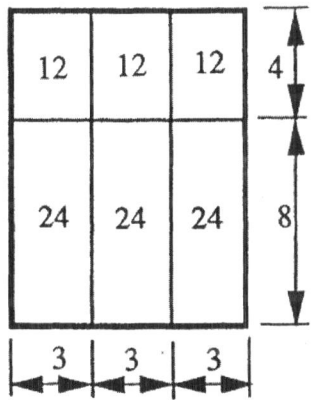

Figure 4. Mullions and transom

The point can be developed if a transom is added to the windows at two-thirds of the height of the mullions, as sketched in Figure 4. The three upper panels of glazing will then be subjected to wind forces of 12 kN, and the lower to 24 kN. Once again an assumption has to be made as to how these forces are carried by the transom and mullions. One possibility, from the point of view of static equilibrium, is that all the wind forces are transferred to the transom (and to the top and bottom frames of the window), so that each of the three portions of the transom is subjected to a uniform (!) load of 18 kN (Figure 5). The transom could be considered as a long member spanning 9 m from side to side, and receiving no support from the mullions; since the transom will be made of short lengths of stone, dowelled together, this is an unreasonable assumption. A much more reasonable view of the action of the transom is that it acts as three flat arches of the type sketched in Figure 2, in which case the equilibrium forces of Figure 5 may be determined.

From this figure it will be seen that each mullion is acted upon by a transverse point load of 18 kN (and by no other wind load) as sketched in Figure 6. The thrust lines in this (cracked) mullion are straight and, for $d/t = 24$ again, the value of H is determined as 96 kN (compared with 108 kN without a transom). In actual fact the transom and mullions will act partially together in some unknown way, but it must be emphasized that the 'reasonable' calculations leading to Figures 5 and 6 are safe. Consideration of further and much more complex sub-framing of the window (additional smaller mullions and further transoms) leads to the same conclusion that the mullions are subject to end forces of about 100 kN.

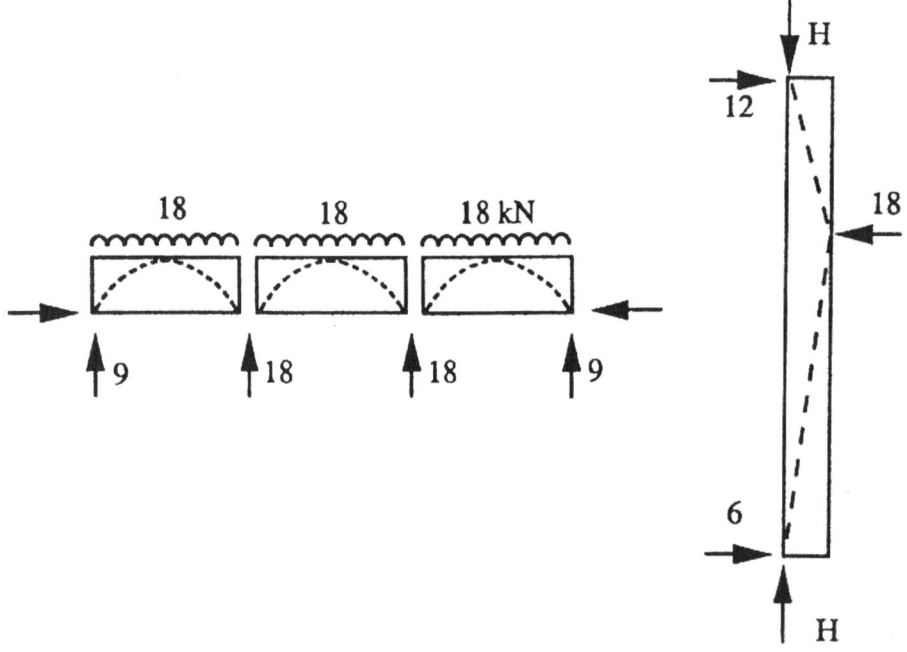

Figure 5 (above). Equilibrium of transom
Figure 6 (right). Equilibrium of mullion

As an example of interactive response of the members of a window, the artificial framing of Figure 7 may be considered. The wheel window shown is supposed to have only 3 'spokes', each of length 2 m, supporting the glazing. As before, the wind load will be taken as 1 kN/m^2; thus the window as a whole is subject to a load of 4π or 12.6 kN. Some of this load may pass directly to the circumference of the window, and some may be carried by the three spokes – a worst case would arise if the *ferramenta* distributed the load only to the spokes, so that each carried 4.2 kN. This load will not be distributed uniformly, but in the absence of information about the disposition of the glazing bars, the equilibrium of a spoke will be considered as sketched in Figure 8. If the ratio d/t has value 10 (i.e., a stone spoke of length 2 m and depth 200 mm) the value of H is determined as 21 kN.

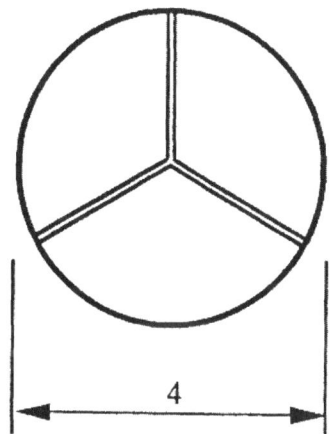

Figure 7. Basic wheel window

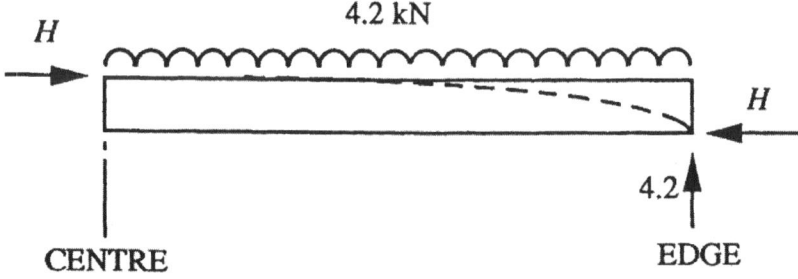

Figure 8. Equilibrium of a spoke

Each of the three spokes thrusts at the centre of the window with a force H (= 21 kN); the three forces lie in a plane and are inclined to each other at 120°, and are therefore self-equilibrating. Figure 9 displays the equilibrium state of the window under the transverse loading of 12.6 kN, and is a simple and artificial but archetypal example of the kind of analysis that is needed for the assessment of the stability of rose windows. The 'spokes' of such a window thrust against the surrounding masonry, and the values of those thrusts (e.g., $H = 21$ kN) must be determined. Further, there must exist flat planes of thrust within the members of the window, even if these members show some moderate curvature. The internal thrusts may of course change their directions at junctions between members. The paths will appear straight in an elevation such as Figure 9; within individual members, the thrusts will follow paths such as that shown by the broken line in Figure 8.

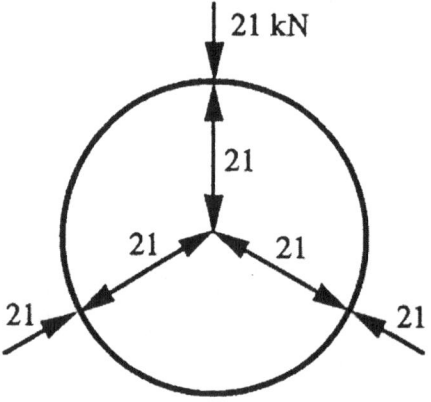

Figure 9. Forces in spokes

Equilibrium of rose windows

A simple wheel window is sketched in Figure 10, and is supposed to be subjected to transverse wind load. It is clear that, once the reactions at the ends of the spokes have been found, a satisfactory pattern of equilibrium forces can be found for the whole window. The end reactions will be transmitted directly down the spokes; at their inner ends, the spokes are shown in Figure 10 as terminating against an inner ring of masonry. The forces from the spokes will put the ring into (almost) uniform circumferential compression; in fact, a regular twelve-sided figure, a dodecagon, may be imagined as contained within a circular ring as recording the traces of the internal compression. Thus the geometry of an arrangement such as that shown in Figure 10 is satisfactory.

In order to check stresses, the end reactions at the spokes must be determined, and a very simple approximate formula may be found. Figure 11 imagines the

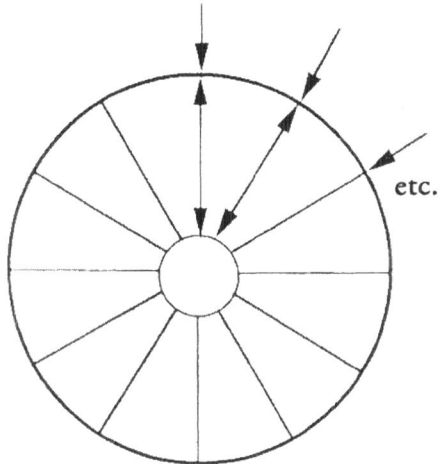

Figure 10. Simple wheel window

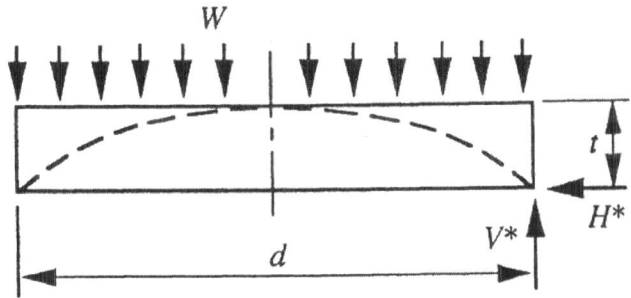

Figure 11. Wheel window with infinitely many spokes

window of Figure 10 to be 'smeared' into a continuous circular slab, where the external wind load W is resisted by the domical thrusts indicated by the broken line. The sum of the horizontal reactions at the perimeter of the window of diameter d must of course equal W; the reaction V^* is supposed to be distributed round the window. Similarly the thrust H^* of the window against its circular frame is also distributed round the circumference. For small values of t/d it is easy to show that

$$H^* = \frac{Wd}{4t} \qquad (2).$$

Thus for the three-spoke example of Figure 7, for which d = 4 m and t = 200 mm, the value of H^* is 5 W, and for W = 12.6 kN (as in the numerical example), H^* = 63 kN. This *total* reaction is provided by 3 spokes, so that the thrust contributed by each spoke is determined as 21 kN, exactly as shown in Figure 9.

A numerical example

The rose window of Figure 12 has diameter 6 m and has 12 'petals'. The thickness of the masonry is supposed uniform at 300 mm. The wind loading is 2 kN/m^2; thus the total load on the window is W = 56.5 kN. The ratio $d/4t$ has value 5, so that H^* from equation (2) totals 283 kN. Since the window has 12 compartments, the reaction at each of the petal tips is 23.6 kN, as shown in Figure 12. Each of the petal tips can be approximated by an equilateral triangle lying within the boundaries of the curved masonry, so that the analysis can be extended to the inner members of the rose, giving the compressive forces marked in Figure 12. Finally, the conditions in the inner dodecagon are as sketched in Figure 13. The force transmitted round the circle has value $^1/_2$(19.2) cosec 15°, or 37.1 kN, and a masonry section 150 x 300 would keep average compressive stresses below 1 N/mm^2.

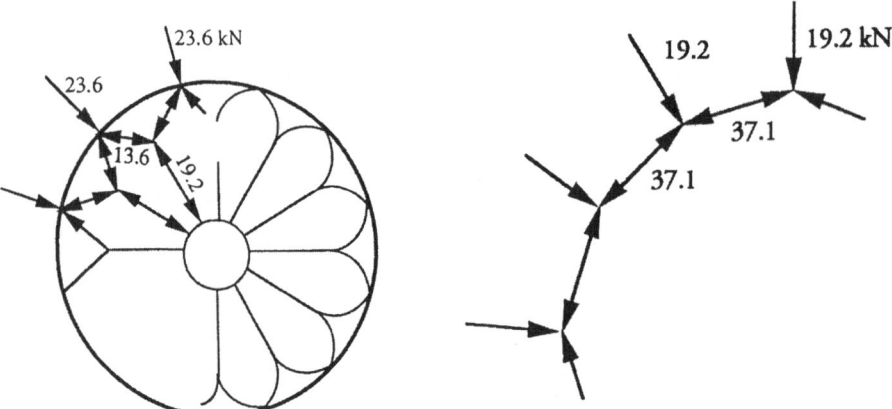

Figure 12. Forces in simple rose windows

Figure 13. Forces in the inner ring

Different notional triangulations are possible for the petal tips, but different assumed angles lead to the same conclusion that loads in the 'spokes' and petal 'triangles' are about 20 kN, and perhaps double this in the central ring (cf. the values marked in Figure 12).

Notre Dame de Mantes

The window at Mantes (c. 1180) sketched in Figure 14 has a diameter of 8 m; all masonry will be taken to have dimensions 250 x 250. At a wind loading of 2 kN/m^2, the total transverse load W is 32π or 100.5 kN. Since $d/4t = 8$, equation (2) gives $H^* = 804$ kN, and each of the 12 outer spokes will thrust against the containing masonry with a force of 67.0 kN. On the left-hand side of Figure 14 is shown a 'reasonable' skeletal structure consisting of straight members containable within the actual curved masonry, and simple triangulation of forces leads to the values marked in the figure. The force round the inner ring (the dodecagon) is calculated as 91.6 kN; average compressive stresses in the members are of the order of 1 N/mm^2.

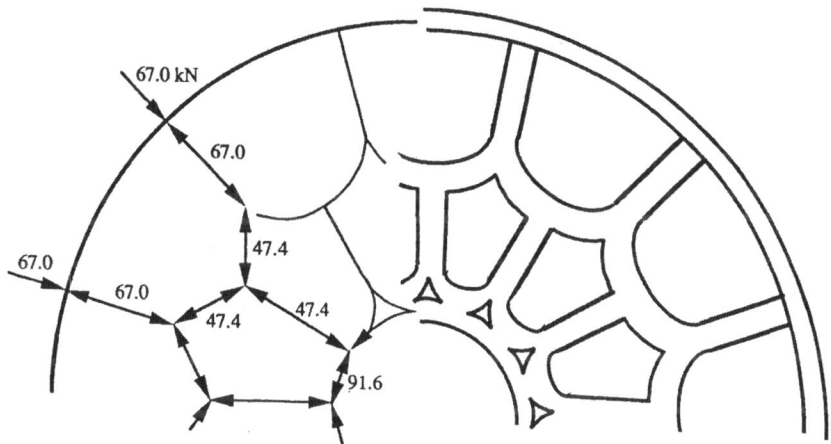

Figure 14. The rose window of Notre Dame de Mantes

The gable end

Figure 15 shows a rose window inserted in a gable end – perhaps to a transept without side aisles, so that the supporting buttresses may be placed against the masonry without the need for flyers. With or without flyers, the buttressing system will, inevitably, give way, and an initially circular window will become oval; the ovality may not be appreciated by the eye, but may in practice amount to a few centimetres out of circular. The framing of the window is represented schematically in Figure 16 by a simple cross; the distortion of the window, greatly exaggerated, will tend to crush the vertical 'spokes', while the horizontal spokes are left poorly supported at their ends.

Figure 15. Rose window in a gable

By contrast, if the framing had incorporated a central ring of masonry – the scheme of Figure 17 – then it is evident that ovalization of the window could be accommodated flexibly, with the central ring also distorting. In Figure 17 the central square has become a lozenge; in Figures 12 and 13 the regular dodecagon contained within the circle would flatten, but remain satisfactory as a load-bearing rib.

As a matter of interest, had the cross of Figure 16 been rotated through 45° (at least one example of such a window exists) then the now diagonal spokes would, to the first order, be unaffected by ovalization of the frame.

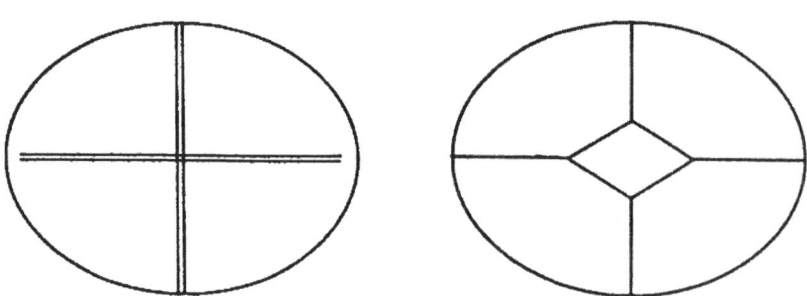

Figures 16 and 17. Ovalisation of a window

Conclusion

The significant loading on a rose window arises from the action of wind. The members of the window are made from relatively small pieces of stone, and cohesion of the assembly must be assured by using connecting dowels, so that transverse forces arising from the wind may be transmitted from stone to stone. Similarly, the peripheral stones of the window must be firmly dowelled to the surrounding masonry.

The transverse wind loading is resisted by flat-arch action of the window as a whole, and a pattern of compressive forces is generated within the components of the window. These components may be curved visually, but they must have geometries such that a compressive 'truss' of straight members may be imagined to be contained within their widths. The members of the window thrust against the surrounding masonry, and they must remain 'tight' in order to transfer the wind loading. In practice the geometry of the opening surrounding the window will distort as a result of shrinkage of mortar, foundation settlements and other unpredictable imperfections, and the rose window itself must have a flexible geometry to accommodate such movements. Such flexibility is inherent in a window in which the spokes are supported by a small inner ring of masonry.

References

COWEN, PAINTON. 1979. *Rose Windows.* London: Thames and Hudson.

HEYMAN, JACQUES. 1972. 'Gothic' Construction in Ancient Greece. *Journal of the Society of Architectural Historians,* vol. 31: 3-9.

————. 1995. *The Stone Skeleton.* Cambridge: Cambridge University Press.

Gaetano Giorgini (1795-1874)

EARLY THEORIES OF VECTORS

Sandro Caparrini[1]

According to a commonly-accepted picture of the development of mathematics, vector calculus arose as a consequence of the discovery of the geometric representation of complex numbers, at the beginning of the nineteenth century. This is not entirely true. In fact, there were some very important early influences from geometry and mechanics, which can be traced back to the works of many mathematicians, notably Euler, Lhuillier, Carnot, Laplace, Poinsot, Poisson, Français, Binet and Cauchy. The two main sources of inspiration lie in the analytic theory of polygons and polyhedra on the one hand, and in the vector representation of moments and angular velocity on the other. As a result, some geometric theories that were roughly equivalent to elementary vector calculus were created by Giorgini, Chasles, Möbius, Saint-Venant and Chelini. At the end of the century they were assimilated into the mainstream of vector calculus.

Introduction

I have never met Prof. Benvenuto, but I was fortunate enough to have several conversations and to correspond with Prof. Truesdell. I was a young graduate in nuclear physics then, and did not know much about mechanics and its history. Probably Truesdell saw clearly my lack of scientific education, but he nevertheless maintained his gentlemanly courtesy toward me. I received much inspiration from his work, and I take this occasion to express my gratitude for his encouragement.

According to every book on the history of mathematics, the birth of vector calculus is a consequence of the discovery of the geometrical interpretation of complex numbers. This widespread conception is easily summarized. Between the years 1799 and 1831 six different mathematicians (Wessel, Argand, Buée, Mourey, Warren and Gauss) independently realized that complex numbers can be represented as directed line segments in the plane, and that they can be added by means of the parallelogram law. In 1844 Hamilton and Grassmann generalised this result, creating, respectively, hypercomplex numbers with four components (quaternions) and exterior algebra. Their theories were greatly simplified to meet the needs of mathematical physics and, in the hands of Heaviside, Gibbs and several other less important mathematicians, became our elementary vector calculus.

While this historical picture appears correct in every detail, a moment's reflection will show that there must be something wrong with it. The main use of vector calculus is to operate directly with physical entities such as forces and

---

[1] Department of Mathematics, University of Turin, via C. Alberto 10, 10123 Turin, ITALY

velocities, and to visualize and simplify the calculations in analytic geometry. Thus it seems quite strange that it bears no relation at all to these two disciplines. This paradoxical opinion was stated explicitly by M.J. Crowe at the beginning of his *History of Vector Analysis*:

> *The idea of a parallelogram of velocities may be found in various ancient Greek authors, and the concept of a parallelogram of forces was not uncommon in the sixteenth and seventeenth centuries. By the early nineteenth century parallelograms of physical entities frequently appeared in treatises, and this usage indirectly led to vector analysis, for this idea provided a striking example of how vectorial entities could be used for physical applications. It should not be inferred, however, that all of those who used the concept of a parallelogram of physical entities were aware of the idea of a vector or of vector addition. The essential idea in the parallelogram of physical entities is the construction of diagram in terms of which the operations involved in determining the resultant become evident. The idea of adding the lines need not be introduced or was it (to my knowledge) ever introduced before the creation of vectors. Thus this idea alone could not and almost certainly did not directly stimulate anyone to the creation of a vectorial system. Its influence was indirect but important, for it was the first and most obvious case in which vectorial methods could be brought to the aid of physical science* [Crowe 1967: chap. I, sec. I].

A somewhat different point of view was expressed by Karin Reich:

> *Within mechanics, an urgent demand for a new calculus seems to have risen, as forces, pairs of forces and the operations with them had to be named and represented. Before Grassmann and Hamilton, many authors of mechanical textbooks chose something similar to vector calculus. Very often there was no clear definition of magnitudes, and only a simplified representation. Examples are Domenico Chelini (1860), Henri Résal, who published many textbooks during the second half of the nineteenth century, and Osip I. Somoff, whose textbook on kinematiks in Russian was translated into German by Alexander Ziwet (1878)* [Reich 1996: 197].

Crowe himself must have felt some uneasiness when he came to describe several vectorial systems that did not seem to lie on the main line of development:

Coincidence is certainly not the explanation for the fact that more than ten men in six countries,[2] in the period from the 1790's to the 1850's, sought to create vectorial systems. Though few of these men knew of the ideas of any of the others, nonetheless some factors in the mathematics and physics of this period must have motivated their search [Crowe 1967: chap. III, sect. II].

My own opinion, corroborated by a wealth of books and articles written by many different mathematicians in the first half of the nineteenth century, is that *there was* a direct influence of mechanics and geometry on the development of vector calculus. Roughly speaking, the scalar product arose from analytic geometry (projection of line segments on straight lines), while the vector product was the translation into the language of pure geometry of the analytic expression for the moment of a force. To demonstrate in every detail such a thesis would require a long essay, which I hope to write soon. Here my scope is more modest, for I want to describe succinctly the work of two mathematicians who constructed mathematical systems in many ways similar to the algebra of vectors, but whose motivations lie outside of the theory of complex numbers. While the present article is only an outline of a more complete exposition, I hope that it will be sufficient to show that there are indeed solid motivations for a new evaluation of the history of vector calculus.

Prologue: the history to 1820

To understand the achievements of some of the mathematicians who created early forms of vector calculus, we must first get a view, albeit an incomplete one, of some results in mechanics and geometry at the beginning of the nineteenth century.

While the geometrical representation of forces and velocities by means of directed line segments (let us call them *vectors*, for this is what they really are) was already fairly well known by the middle of the eighteenth century, moments of forces and angular velocities had to wait for the second half of century to be recognized as vectorial entities. This fact is not really surprising, for they are mainly employed as theoretical tools in the study of the motion of rigid bodies, which was developed in a general form from 1749 onward.[3] I have recounted in detail the history of this important advance in mechanics in a recent paper [Caparrini 2002], and thus need not repeat myself here. That moments of forces are vectors was first discovered by Euler (1780), and rediscovered independently

---

[2] Mainly Möbius, Bellavitis, Hamilton, Grassmann, de Saint-Venant, Cauchy and O'Brien.

[3] Beginning with the publication of d'Alembert's *Recherches sur la Précession des Équinoxes* (1749) and Euler's *Découverte d'un nouveau Principe de Mécanique* (1752).

by Poinsot (1803), while Laplace had come very close to finding this important result with his concept of the invariable plane (1798). The vectorial composition of small rotations had been meanwhile discovered by Frisi (1758), but Lagrange arrived at this theorem in his own way in the second edition of the *Mécanique analytique* (1811) and made it generally known. Thus we see that by 1811 it was clear that the most fundamental objects in mechanics are vectors.

A different approach to the geometrical representation of moments was taken by Poisson. He remarked that the moment of a force about a point is numerically equal to the double of the area of a triangle having the vertex in the point and the force itself as its basis. Thus, implicitly, he assumed that the moment of the force can be represented geometrically by this triangle. Poisson was therefore led to study analytically the projections of plane surfaces (1808).

During the same time that the vectorial theory of mechanics was developed, there grew also a new approach to the theory of polygons and polyhedra. Carnot (1803) and Lhuillier (from 1789 to 1828) studied geometrically the projections of plane surfaces and line segments, while Livet (1806), Jacques Frédéric Français (1808) and Hachette (1809 and 1811) discussed the transformation of non-rectangular Cartesian axes. Their results were united with those of Poisson and expounded by Hachette in his influential textbooks of analytic geometry (1813, 1817).

Having briefly reviewed the state of the knowledge in 1820, we can now consider some remarkable mathematical synthesis of these results.

Gaetano Giorgini (1795-1874)

The Italian mathematician Gaetano Giorgini was born at Montignoso (Lucca) of prosperous parents, so that his education presented no problems. He was sent to Paris, where he attended first a *Lycée*, then the *École polytechnique*. There he came in contact with Monge, Poisson, Arago and especially Chasles, with whom he maintained a lifelong friendship. He showed early signs of promise, for he won a prize for mathematics while still a high school student. When he returned to Italy, he began a career as a civil engineer. Giorgini wrote a few mathematical works, but after about 1835 he gave up the pursuit of mathematics. When he died he was *Senatore del Regno d'Italia*.[4]

Giorgini's reputation today rests upon a single mathematical result, the anticipation of Chasles's and Möbius's *Nullsystem* [see Loria 1893: 26], but his

---

[4] For a complete account of his life and works see [Loria 1893]. Loria does not mention two early articles in his bibliography of Giorgini's writings; they are [Giorgini 1813] and [Giorgini 1814].

most significant work is a long essay, the *Teoria analitica delle projezioni* [Giorgini 1820]. Here he gave a systematic theory of projections for line segments and plane surfaces, in which all the previous results are correlated and generalized. Undoubtedly, the motive that led Giorgini to write this small treatise was the desire to complete the work of his predecessors, but the result bears an uncanny resemblance to an exposition of our elementary vector algebra, expressed in components. This is not a coincidence; in fact, Giorgini is studying the elements of the affine group of transformation of coordinates in space. The main differences between his formulae and those reported in textbooks of today are due to the greater scope and generality of his approach, which is not limited to orthogonal transformations.

The *Teoria analitica* was first printed as a small booklet of sixty-eight pages, then reprinted in an Italian mathematical journal.[5] It is divided into six parts. The first chapter treats the projections of line segments in two dimensions, the second chapter extends the results to three dimensions. Chapter III considers the projections of plane surfaces in space. In the fourth chapter Giorgini studies the projections of three-dimensional objects formed by line segments and plane surfaces. The last two chapters are dedicated to the applications of the theory developed so far to pure geometry and mechanics. There is no introduction or preamble, so that we do not know how Giorgini viewed his own theory. However, he cites the names of many mathematicians who had worked on the composition of moments and the theory of projections: Hachette, Binet, de Prony, Poisson, Laplace. Presumably, he considered his theory a generalization to non-orthogonal axes of the chapter on the projections of line segments and surfaces given by Hachette in his textbook of analytic geometry, entitled *Théorie des surfaces* [Hachette 1813: chap. I, sec. III].[6]

The style of the work is for the most part purely analytic, the geometric considerations being reserved for the demonstration of the basic formulae. It is a paper dense with results, each of which is obtained by means of relatively easy proofs. Almost all of the results are original, or are generalizations of old ones.

Giorgini begins his essay by introducing the two kind of projections of a line segment a on the axes of a non-orthogonal system of Cartesian coordinates: the parallel projections (*projezione oblique*), denoted by a_x, a_y, a_z, and the orthogonal projections (*projezioni ortogonali*), denoted by $\bar{a}_x$, $\bar{a}_y$, $\bar{a}_z$. In the language of tensor calculus the parallel projections and the orthogonal

---

[5] In the present account of this work, I use the edition published in the *Atti*. On the title page of this edition it is said that the work was read before the Academy of Lucca on 3 June 1819.

[6] Hachette published a second, much expanded version of the same book under the new title *Éléments de Géométrie à trois dimensions* [Hachette 1817].

projections are now called respectively the *contravariant* and the *covariant* components of the vector **a**, and are denoted by a^i and a_i.[7] Thus we see that even in matters of notation Giorgini is not much removed from us. Moreover, it must be noted that this is one of the first clear-cut distinctions between the two types of projections in analytic geometry.[8]

In summarizing the first part of the memoir, we confine our attention to the analysis in three dimensions. In fact, the first chapter seems redundant, for in the second chapter we have a more satisfactory treatment of the subject. However, we note that the first chapter contains a detailed description of the signs that must be assigned to the projections of a line segment. Giorgini considers a line segment in the plane, referred to a non-orthogonal coordinate system, and studies the signs of his projections on the coordinate axes for every orientation of the segment. At the end of his analysis he says:

> *We have given these peculiarities in every detail, but we have been constrained to do so because of the importance of the consideration of signs in making use of projections.*[9]

Statements of this kind show that Giorgini had in mind directed segments and oriented surfaces, that is, a rule for signs of geometrical objects akin to that ordinarily used today.

The results given in chapter II are entirely based on a single theorem, which furnishes the angle between two directed line segments *a* and *a'*. Giorgini's formula [1820: 48] is

$$\frac{\sin a'.X}{\sin x.X}\cos a.x + \frac{\sin a'.Y}{\sin y.Y}\cos a.y + \frac{\sin a'.Z}{\sin z.Z}\cos a.z = \cos a.a' \,,$$

where *a.X, a.Y, a.Z* are the angles formed by the segment *a* respectively with the coordinate planes *yz, xz, xy*. The proof is accomplished by showing that the projection of a directed segment along a given line can be obtained by adding the projections of its components with respect to the coordinate axes. In modern terms, the scalar product of a vector **a** for another vector b can be obtained by summing the scalar products of its components:

$$\mathbf{a}\cdot\mathbf{b} = \left(\mathbf{a_x}+\mathbf{a_y}+\mathbf{a_z}\right)\cdot\mathbf{b} = \mathbf{a_x}\cdot\mathbf{b}+\mathbf{a_y}\cdot\mathbf{b}+\mathbf{a_z}\cdot\mathbf{b}$$

---

[7] For a proof of this assertion see, for example, Pauli [1958: sec. 10].

[8] They had been already defined and used by Hachette [1809: 7].

[9] *Abbiamo dato queste particolarità alquanto minutamente, ma siamo stati indotti a ciò fare per l'importanza delle considerazioni dei segni nel fare uso delle projezioni* [Giorgini 1820: 35].

21

$$(8)\ldots \frac{sena'.X}{sen x X}\,\overline{a}_x + \frac{sena'.Y}{seny.Y}\,\overline{a}_y + \frac{sena'.Z}{sen z.Z}\,\overline{a}_z = \overline{a}_{a'}\,,$$

$$(9)\ldots a'_x \cos a.x + a'_y \cos a.y + a'_z \cos a.z = \overline{a'}_a\,,$$

le quali equazioni risolvono i due problemi seguenti.

Problema 11.° » Date le tre projezioni orto-
» gonali di una retta sopra tre assi obliqui, de-
» terminare la projezione ortogonale della me-
» desima sopra una retta data. »

Problema 12:° » Date le tre projezioni oblique
» di una retta, determinare la projezione ortogo-
» nale della medesima sopra una retta data »

L'equazioni (7) moltiplicate per a divengono, mediante le formole (1) e (2),

$$(10)\ldots \begin{cases} \ldots a_x + a_y \cos x.y + a_z \cos x.z = \overline{a}_x\,, \\ \ldots a_y + a_x \cos x.y + a_z \cos y.z = \overline{a}_y\,, \\ \ldots a_z + a_x \cos x.z + a_y \cos y.z = \overline{a}_z\,, \end{cases}$$

e risolvono il

Problema 13.° » Date le projezioni oblique di
» una retta, determinarne le projezioni ortogona-
» li e reciprocamente ».

Nella categoria degli antecedenti problemi, ma più generale, è il seguente

Problema 14.° » Date le tre projezioni di una
» retta sopra tre assi coordinati, determinare le
» tre projezioni della medesima sopra tre nuovi
» assi »

Per risolvere questo problema, si rappresenti-
no per $x', y', z'; Z', Y', X'$ gli assi ed i piani coordinati del nuovo sistema, si cambi nell'equazione (5) a in a', ciò che la trasforma nella se-
guente

Figure 1. Giorgini's expression for the relation between the covariant and the contravariant components of a vector a. Page 21 from Giorgini's *Teoria analitica delle projezioni*, 1820

This demonstration had been first used by Hachette [1809: 8], but Giorgini does not refer to him.

We remark that the notation (a, b) for the angles between two segments a and b had been introduced by Français [1808: 182], and generalized by Hachette [1811: 248] to include the angles between straight lines and planes. This notation is very useful for showing some striking similarities between different formulae, as we will see in what follows.

Giorgini demonstrates easily the following formulae [1820: 46, 48]:

$$a_x = a\frac{\sin a.X}{\sin x.X} \; , \quad a_y = a\frac{\sin a.Y}{\sin y.Y} \; , \quad a_z = a\frac{\sin a.Z}{\sin z.Z} \; ,$$

$$\bar{a}_x = \cos a.x \, , \quad \bar{a}_y = a\cos a.z \, , \quad \bar{a}_z = a\cos a.z \, , \quad \bar{a}_b = a\cos a.b \, ,$$

which express the parallel and the orthogonal projections by means of the length of the segment. Substituting in the preceding equation, he finds the relations between the two kind of projections (Figure 1) [Giorgini 1820: 49]:

$$a_x + a_y \cos x.y + a_z \cos x.y = \bar{a}_x \, ,$$

$$a_y + a_x \cos x.y + a_z \cos y.z = \bar{a}_y \, ,$$

$$a_z + a_x \cos x.z + a_y \cos y.z = \bar{a}_z$$

In modern terms, he has found, in a particular case, the well-known formula of tensor calculus for the lowering of indices. This is the first appearance of this relation, which now plays an important part in differential geometry and mathematical physics. Let g_{ik} denote the metric tensor, a_i and a^i respectively the covariant and the contravariant components of a vector a; then

$$g_{ik} a^k = a_i \, , \quad i = 1, 2, 3,$$

where we have used the Einstein convention for the indices. Obviously, Giorgini considers only the particular case in which

$$g_{11} = g_{11} = g_{11} = 1, \; g_{12} = g_{21} = \; \cos x.y \, , \; g_{23} = g_{32} = \; \cos y.z \, , \; g_{13} = g_{31} = \; \cos x.z.$$

Next Giorgini finds the relations between the components of a directed line segment in two different non-orthogonal systems of coordinates. He remarks that they include as a special case the usual formulae for the transformation of axes [Giorgini 1820: 49-50].

Giorgini states also something more in the next few pages. Here we can find for the first time several expressions for the *scalar product* of two vectors. Giorgini writes the following formulae [Giorgini 1820: 51]:

» sopra lo stesso asse, è uguale al prodotto del-
» le due rette moltiplicato nel coseno dell' angolo
» compreso . »

In forza di questo teorema avremo parimente

$$(13)\ldots a_x \overline{a'}_x + a_y \overline{a'}_y + a_z \overline{a'}_z = aa' \cos a.a'$$

e quindi

$$a'_x \overline{a}_x + a'_y \overline{a}_y + a'_z \overline{a}_z = a_x \overline{a'}_x + a_y \overline{a'}_y + a_z \overline{a'}_z,$$

risultato per se stesso osservabile .

L' equazione (6) moltiplicata per a^2 diviene

$$(14)\ldots a_x \overline{a}_x + a_y \overline{a}_y + a_z \overline{a}_z = a^2,$$

Teorema 17.° » La somma de' tre prodotti di
» ciascheduna projezione obliqua per la projezio-
» ne ortogonale corrispondente di una stessa li-
» nea è uguale al quadrato della retta medesima . »

Nell' equazioni (12) e (14) sostituiti i valori
(10), si ottengono le seguenti

$$(15)\ldots a_x a'_x + a_y a'_y + a_z a'_z + \left(a_x a'_y + a_y a'_x \right) \cos x.y$$
$$+ \left(a_x a'_z + a_z a'_x \right) \cos x.z + \left(a_y a'_z + a_z a'_y \right) \cos y.z \Big\} = aa' \cos a.a,$$

$$(16)\ldots a^2_x + a^2_y + a^2_z + 2a_x a_y \cos x.y + 2a_x a_z \cos x.z$$
$$+ 2a_y a_z \cos y.z = a^2;$$

risultati osservabili, il secondo dei quali sommini-
stra il quadrato di una retta in funzione delle sue
projezioni oblique e coincide colla nota espressione
della diagonale un parallelepipedo, allorquando la
retta a, essendo condotta per l' origine, può esser
considerata come diagonale di un parallelepipedo di cui a_x, a_y, a_z sono i tre lati .

13.° Si consideri attualmente un sistema di
rette a, a', a'' ec. di grandezza determinata e
poste a volontà nello spazio; si rappresentino
inoltre per A_x, A_y, A_z le tre somme

Figure 2. Different forms of Giorgini's scalar product. Page 23 from Giorgini's *Teoria analitica delle projezioni*, 1820.

$$a_x b_x + a_y b_y + a_z b_z + \left(a_x b_y + a_y b_x\right)\cos\,x.y + \left(a_x b_z + a_z b_x\right)\cos\,x.z$$
$$+ \left(a_y b_z + a_z b_y\right)\cos\,y.z = ab\,\cos\,a.b$$
$$a_x \overline{b}_x + a_y \overline{b}_y + a_z \overline{b}_z = ab\cos a.b\,.$$

It is easy to translate them into modern language. Let g_{ik} be the metric tensor considered above; then:

$$a^i b_i = \mathbf{a}\cdot\mathbf{b}\,,\;\; g_{ik}a^i b^k = \mathbf{a}\cdot\mathbf{b}\;\;(i,k=1,2,3)\,.$$

Giorgini does not attach any particular meaning to these formulae. He seems to regard them simply as pleasantly symmetric expressions concerning the projections of line segments (Figure 2). The second half of the chapter is concerned with the generalization of the previous formulae to systems of directed segments. By considering the algebraic sum of the projections of several segments on the three coordinate axes, he tacitly introduces into his theory the *composition of directed segments by the parallelogram law*. Giorgini draws attention to the fact that his formulae generalize to non-orthogonal axes those given by Hachette in his *Traité des Surfaces* [1813].

In chapter III Giorgini carries out for plane surfaces the program that he had previously attained for directed segments in space – their study through projections on the coordinate planes. He obtains a series of formulae remarkably similar to those for segments, so that they could be put into one-to-one correspondence by a simple interchange of symbols, thus establishing a kind of 'principle of duality' for projections. In fact, he says:

> We will show in the following paragraphs that the projections of plane surfaces enjoy the same properties as those already demonstrated for the projections of line segments.[10]

Today the meaning of such duality is easy to understand. We now know that it is possible to represent a plane surface vectorially by taking a vector whose direction is perpendicular to the surface and whose length proportional to its area. In fact, the vector product *a×b* is simply the vectorial representation of the parallelogram constructed on the vectors *a* and *b*. Therefore, when Giorgini considers the projections of the surfaces on the coordinate planes, he is implicitly using the components of a vector. In other words, he is studying a mathematical operation almost equivalent to the vector product.

---

[10] *Faremo vedere nei susseguenti articoli che le projezioni delle aree piane godono di proprietà analoghe a quelle già dimostrate per le projezioni delle rette* [Giorgini 1820: 59].

10

e moltiplicandole per a, facciamo uso delle formole (1), e delle seguenti che ne derivano

$$a_{x'} \operatorname{sen} x'.y' = a \operatorname{sen} a.y' ,$$

$$a_{y'} \operatorname{sen} x'.y' = a \operatorname{sen} a.x' ,$$

ciò che darà per risolvere il problema l'equazioni

$$(13) \ldots \begin{cases} a_x \operatorname{sen} x'.x - a_y \operatorname{sen} x'.y = a_{y'} \operatorname{sen} x'.y' , \\ a_x \operatorname{sen} y'.x - a_y \operatorname{sen} y'.y = a_{x'} \operatorname{sen} x'.y' . \end{cases}$$

L'equazione (5) moltiplicata per aa' diviene

$$(14) \ldots a'_x \overline{a}_x + a'_y \overline{a}_y = aa' \cos a.a'$$

e dimostra il

Teorema 3.° » La somma de' due prodotti di
» ciascheduna projezione obliqua di una retta per
» la projezione ortogonale corrispondente di un al-
» tra retta è uguale al prodotto delle due rette
» moltiplicato pel coseno dell' angolo compreso. »

L'equazione (7) moltiplicata per aa' dimostra il

Teorema 4.° » Tra due rette a ed a' e le lo-
» ro projezioni ortogonali sopra i due assi conju-
» gati si verifica l' equazione

$$\text{» } (15) \ldots \frac{1}{\operatorname{sen} x.y} \left(\overline{a}_y \overline{a'}_x - \overline{a}_x \overline{a'}_y \right) = aa' \operatorname{sen} a.a' \text{ »}$$

Similmente l' equazione (8) moltiplicata per
aa' dimostra il

Teorema 5.° » Tra due rette a ed a' e le lo-
» ro projezioni oblique si verifica l' equazione

$$\text{» } (16) \ldots \left(a_y a'_x - a_x a'_y \right) \operatorname{sen} x.y = aa' \operatorname{sen} a.a' \text{ »}$$

l' equazione (6) moltiplicata per a^2 diviene

$$(17) \ldots a_x \overline{a}_x + a_y \overline{a}_y = a^2 ,$$

e dimostra il

Figure 3. The two-dimensional vector product as stated by Giorgini. Page 10 from Giorgini's *Teoria analitica delle projezioni*, 1820.

To show the extent of this duality, we consider the equations that correspond to those given above for the transformation of coordinates [Giorgini 1820: 62]:

$$m_X + m_Y \cos X.Y + m_Z \cos X.Z = \overline{m}_X \,,$$

$$m_Y + m_X \cos X.Y + m_X \cos Y.Z = \overline{m}_Y \,,$$

$$m_Z + m_X \cos X.Z + m_Y \cos Y.Z = \overline{m}_Z \,,$$

where m_X, m_Y, m_Z, and $\overline{m}_X$, $\overline{m}_Y$, $\overline{m}_Z$, are respectively the parallel projections and the orthogonal projections on the three coordinate planes of the plane area m. The formal identity between these equations and those obtained for the projections of segments is immediately apparent.

One notes with surprise that Giorgini does not take advantage of the geometric representation of plane surfaces by means of a normal vector. The possibility of such a representation is quite straightforward from his equations, but he does not seem to see its meaning and importance. This is the main departure of Giorgini's theory from vector calculus.[11]

The sum of directed segments has here its analogue for plane surfaces. Giorgini considers the sum of the projections of a system of plane surfaces on the three coordinate planes, thus creating implicitly a new geometric operation: *the sum of plane surfaces*. Giorgini gives here an almost complete treatment of this concept. It was not really new, for it had been already introduced by Poisson in 1808 [see Caparrini 2002: sec. 8]. Poisson had tried to generalize the ideas behind the theorem of the 'conservations of areas', as the conservation of moment of momentum was then called.

While Giorgini's projections of plane surfaces are not completely equivalent to our vector product, we note with some astonishment that in the first chapter he studies the product $ab \sin a.b$, where a and b are the length of two line segments. He obtains the formulae [Giorgini 1820: 58]:

$$\frac{1}{\sin x.y}\left(\overline{a}_y \overline{b}_x - \overline{a}_x \overline{b}_y\right) = ab \sin a.b \,, \quad \left(a_y b_x - a_x b_y\right)\sin xy = ab \sin ab \,,$$

which are clearly the analytic expressions for the modulus of the two-dimensional *vector product* in non-orthogonal coordinates (Figure 3).

Chapter IV is a short one (it is only three pages long), yet it contains another result of great significance. By considering the angles between segments and

---

[11] The idea of orienting a plane surface by means of the direction of its normal had been first introduced by Poisson in the first edition of his *Traité de Mécanique* [Poisson 1811, 1: 101].

plane surfaces Giorgini foreshadows the *mixed product* of vectors. (As we shall see in the next section, this possibility was later exploited by Chasles.) Giorgini's main result here are the following formulae [Giorgini 1820: 69]:

$$a_x m_X \sin x.X + a_y m_Y \sin y.Y + a_z m_Z \sin z : Z = am \sin a.m \, ,$$

$$\frac{1}{\sin x.X} \bar{a}_x \bar{m}_X + \frac{1}{\sin y.Y} \bar{a}_y \bar{m}_Y + \frac{1}{\sin z.Z} \bar{a}_z \bar{m}_Z = am \sin a.m \, ,$$

where m is a plane surface and a is a line segment. When the axes are orthogonal, they are both reduced to the expression [Giorgini 1820:71]:

$$a_x m_X + a_y m_Y + a_z m_Z = am \sin a..m.$$

It is easy to see that this is the scalar product of a segment and a plane area.

The paper concludes with two chapters on the applications to mechanics and geometry, which do not fall within the scope of the present study.

The *Teoria analitica* does not exhaust the contributions of Giorgini to the theory of projections. In 1828 he published a paper entitled *Sopra alcune proprietà de'piani de'momenti principali e delle coppie di forze equivalenti* [Giorgini 1828], where he made use of projections to demonstrate some theorems in statics but did not introduce any new principles.

To Giorgini belongs the merit of having created a mathematical theory that contains most of the results of our algebra of vectors. However, his *Teoria analitica delle projezioni* went virtually unnoticed, probably because it was written in Italian, and all of its results had to be rediscovered later.

Michel Chasles (1793-1880)

The work of Giorgini on the theory of projections was continued by the great French geometer Michel Chasles, a decade after the publication of the *Teoria analitica delle projezioni*. Chasles is commonly regarded as one of the great geometers of his century, but there is one aspect of his work that is not well known: he showed a deep interest in the early applications of vectors to mechanics and geometry. Thus, for example, in his historical treatises *Aperçu historique sur l'origine et le développement des méthodes en géométrie* [1837] and *Rapport sur les progrès de la géométrie* [1870] he called attention to the importance of Poinsot's results in mechanics for pure geometry.

Chasles collected his ideas on this subject in an article entitled *Mémoire de géométrie pure, sur les systèmes de forces, et les systèmes d'aires planes; et sur les polygones, les polyèdres, et les centres des moyennes distances* [1830]. The title indicates more than adequately its contents, for it says that methods of 'pure geometry' will be used to elucidate questions on 'systems of forces'. Thus the reader is led to expect a work on the geometrical representation of forces. In fact, it turns out to be an exposition of a primitive version of vector calculus.

While the work of Chasles overlaps with that of Giorgini in subject matter, Chasles proved his theorems mainly by synthetic methods and taking full advantage of the representation of plane surfaces by means of vectors. Chasles's paper may be considered the geometric counterpart of Giorgini's essay. He refers to the *Teoria analitica* as "un excellent écrit sur la théorie analytique des projections" [Chasles 1830: 102], thus making it clear that he knows it very well.[12] Moreover, while Giorgini's essay was essentially a work in analytic geometry, Chasles's *Mémoire* is a theory of the geometric composition of segments and plane surfaces. Chasles's own formulation is given in a particular geometrical form, but it is substantially equivalent to our geometric theory of vectors.

The memoir is divided into two parts. In the first part Chasles gives the general theorems on the projections of segments and surfaces, while in the second part, following the example of Giorgini, he applies these results to polygons, polyhedra and to the *centre des moyennes distances.*[13]

The first chapter expounds some results in the theory of the composition of forces acting on a rigid body. It constitutes the basis of the entire work, and thus we will discuss it in detail.

Chasles's theory rests entirely on a single result, which is enunciated thus:

Théorème I. Quand on a deux systèmes de forces, si on multiplie chaque force du premier système par chaque force du second système, et par le cosinus de l'angle de ces deux forces, la somme de tous ces produits sera la même que la somme des produits semblablement faits à l'égard de deux autres systèmes de forces équivalens respectivement aux deux proposées [Chasles 1830, p. 92].

In modern terminology, let a_1, a_2, a_3, ... etc., and b_1, b_2, b_3, ... etc., be two equivalent systems of forces. They are equal to the resultant R; hence, by scalar multiplication, we obtain

---

[12] In a footnote to his paper Chasles writes: "M. Giorgini de son côté, était parvenu précédemment à ces théorèmes. Lui ayant communiqué l'an dernier plusieurs propriétés générales des systèmes de deux forces équivalens, objet dont il venait de s'occuper pour faire suite à sa théorie des projections, j'eus la satisfaction d'apprendre, qu'ainsi que cela nous était arrivé souvent dans nos premières études, nous nous étions rencontrés dans la plupart des résultats, quoiqu'ayant suivi deux marches différentes" [Chasles 1830: 112].

[13] The *centre des moyennes distances* was simply the centre of mass defined in purely geometric terms. It had been introduced by L. Carnot in his *Géométrie de position* [Carnot 1803: 315].

$$R^2 = (\mathbf{a}_1 + \mathbf{a}_2 + \mathbf{a}_3 + ...)(\mathbf{b}_1 + \mathbf{b}_2 + \mathbf{b}_3 + ...) = \sum_i \sum_k a_i b_k \cos(\mathbf{a}_i \cdot \mathbf{b}_k)$$

and the product remains the same for every couple of equivalent systems of forces. This result, which is obvious in modern vector calculus, was entirely new at that time. Chasles's theorem is easily seen to be equivalent to the scalar product of vectors and to include its most important features, such as the distributive and the commutative properties. Chasles is aware of the generality of this result:

> La formule Σ a.b cos.(a.b) = const. qui représente le théorème que nous venons de démontrer, est la base unique de tout cet écrit: elle est susceptible d'autres interpretations géométriques différentes, qui offrent des principes tout aussi généraux que les précédent. Ces principes donnent lieu à un grand nombre de conséquences dont nous ne présenterons dans ce moment que celles de géométrie pure, remettant à un autre écrit celles qui renferment des formules de géométrie analitique [Chasles 1830; 93-94].

Chasles gives some special cases of his general result. The particular case in which the two systems of forces coincide deserves to be noticed, for in this case he obtains the expressions,

$$A^2 = \sum a^2 - 2 \sum a.a' \cos(a, a')\ \text{[Chasles 1830: 94]},$$

$$M^2 = \sum m^2 - 2 \sum m.m' \cos(m, m')\ \text{[Chasles 1830: 102]},$$

where a, a', ... are forces and A is their resultant, m, m', ... are moments of forces and M is the total moment. They had been demonstrated previously by Binet [1815: 326, 337] and Giorgini [1820: 90].

One of the most interesting features of Chasles's work is that his purely geometric theorems are introduced under the travesty of results in mechanics. In fact, the request that the systems of forces satisfy the conditions of equilibrium of a rigid body are *sufficient* for the validity of the theorem, but are not *necessary*. The principle involved in the demonstration is only that the resultant remains constant as the systems are changed; there is no need to require also that the resultant moment should be the same for every system. The artificiality of this point of view was fairly well recognized by Chasles. After deriving the first consequences of his fundamental theorem, he makes the following remark:

> Les forces étant représentées en grandeur et en direction par des droites, on pourrait substituer dans l'énoncé du théorème I le mot droite au mot force. Alors on entendrait par composantes d'une

droite ses projections sur trois axes quelconques menés par un de ses
points; et par systèmes de droites équivalens, *deux systèmes de*
droites dont l'un serait formé par la décomposition et la
composition des droites de l'autre système, comme si ces droites
étaient des forces [Chasles 1830: 94].

The results obtained by Chasles are, in truly vectorial fashion, completely intrinsic, that is, independent of the particular frame in use. However, to prove his theorems he usually introduces a system of non-orthogonal Cartesian coordinates.

The second chapter concerns the composition of plane surfaces. It begins right off with a geometrical proposition:

Lemme. Si l'on projette une aire plane sur trois plans coordonnés
quelconques yz , xz et xy , et si l'on décompose une droite, de
grandeur donnée, et perpendiculaire au plan de cette aire, en trois
autres dirigées suivant trois axes Ox', Oy' , Oz' , respectivement
perpendiculaires aux trois plans yz , xz et xy , les composantes de
cette droite seront aux projections de l'aire plane comme la droite
elle-même sera à cette aire [Chasles 1830: 95].

The lemma asserts, in effect, that we may describe the orientation and the area of a plane surface by considering a vector orthogonal to it. We have here the first appearance of *the representation of plane surfaces by means of vectors*. While this geometric operation had been already adumbrated in the works of Poinsot and Poisson and was implicit in Giorgini's equations, the merit of its introduction should be ascribed to Chasles.

It is evident that the use of the normal vector makes it possible to introduce explicitly the composition of surfaces according to the parallelogram law. Thus, after several near misses, this important geometric operation finally appears:

Corollaire. Il résulte de ce principe que si l'on a un système d'aires
planes situées dans des plans quelconques, et qu'on conçoive un
système de droites perpendiculaires à ces plans et proportionelles
aux aires respectivement, puis qu'on décompose ces droites, de
manière à former un autre système équivalent, et qu'on conçoive
dans des plans perpendiculaires à ces droites des aires qui soient
avec elles respectivement, dans le rapport des aires proposées aux
premières droites, on aura un nouveau système d'aires qu'on aurait
pu former avec le système proposé par des projections des aires de
ce système, analogues aux décompositions des premières droites.
Nous dirons par cette raison que ce nouveau système d'aires est
équivalent au au système d'aires proposé; et l'aire qui correspondra

à la résultante de toutes les droites sera l'aire résultante de toutes les aires; cette aire seule représentera un système équivalent au système des aires proposées. Nous appellerons composantes d'une aire ses projections sur trois plans quelconques, de même que nous appelons composantes d'une droite ses projections sur trois axes [Chasles 1830: 95-96].

Chasles now adds a remark on the work of his predecessors:

Le lemme précédent n'est au fond que le principe dont M. Poinsot s'est servi pour la composition et décomposition des couples. Envisagé d'une manière générale, par rapport à des aires planes quelconques, ce principe présente comme tout-à-fait évidentes des vérités qui ont été longtemps le sujet de beaux théorèmes sur la projection des aires planes et sur la composition des momens. (Voyez Géométrie de position de M. Carnot, la Statique *de M. Poinsot, la* Mécanique *de M. Poisson, et les* Élémens de géométrie à trois dimensions *de M. Hachette)* [Chasles 1830: 96].

Starting from his theorems on the composition of plane surfaces, Chasles goes back to the problems in mechanics which had been studied by Laplace, Poinsot and Poisson and treats them anew. In this connection, he says:

En effet, M. Poinsot a appris, en créant la théorie des couples, à les décomposer comme nous avons décomposé les aires planes, et à les représenter par des droites proportionelles à leurs énergies, et perpendiculaires à leurs plans. On peut donc substituer dans le théorème II, au mot aire, celui de couple. Ainsi le théorème est démontré [Chasles 1830: 99].

This remark shows clearly how the mathematicians of this period took the geometric representations of moments as the model for the creation of a purely mathematical operation, which in time became our vector product.

Mention should also be made of the word *énergie*, which appears in the citation. In the first half of the nineteenth century it was used by some mathematicians (for example, Poinsot and Biot) to denote the 'intensity' of a vector quantity. This usage has never been cited by the many historians of science who studied the different stages of development of the energy concept.

Using his results, Chasles is able to derive the well-known theorems on the composition of moments and Laplace's theory of the invariable plane [Chasles 1830: 99-102].

The third chapter treats the composition of directed segments and plane surfaces. It corresponds to chapter IV of Giorgini's *Teoria analitica*. Like the previous work, it concerns in effect some results which would be best expressed by means of the modern mixed product of vectors:

> *Théorème V. Quand on a un système de droites et un système d'aires, si l'on fait le produit de chaque droite par chaque aire et par le sinus de l'angle que cette droite fait avec le plan de cette aire, la somme des produits ainsi formés aura une valeur qui restera constante quand on substituera au système de droites et au système d'aires, deux autres systèmes de droites et d'aires qui leur soient équivalens* [Chasles 1830: 103].

Obviously the products to which Chasles refers are in effect the volumes of the polyhedra delimited by the plane surfaces and the segments. Here Chasles introduces a concept that had escaped Giorgini, for he considers *oriented* volumes [Chasles 1830: 106]. Moreover, he demonstrates a theorem that is sometimes attributed to him: if we reduce in any way the system of forces acting on a rigid body to two forces, the tetrahedra formed by joining their four ends have a constant volume [Chasles 1830: 109].[14]

The second part of Chasles's paper concerns the applications of his theory to pure geometry, and will not be discussed here.

In the following years Chasles did not develop his theory further, but returned to the same subject in a paper entitled *Théorèmes généraux sur les systèmes de forces et leurs moments* [1847]. Curiously enough, this time he took out of his exposition every reference to the theory of projections, and gave only the theorems on mechanics and geometry. We may conjecture that Chasles did not want to see his old theory compared with the new forms of geometric calculus that had been meanwhile put forward by several mathematicians, such as Cauchy's *clefs algébriques* and Saint-Venant's theory of the *sommes géométriques*. The appearance of Chasles's primitive formulation of geometric calculus would have been quite odd by then, when its most important ideas had been completely reformulated and greatly improved.

Our analysis of Chasles's memoir shows that it deserves to be put alongside the works of the creators of vector calculus. It is difficult to estimate its influence on later writers, for it does not seem to have met any immediate response. However, taking into account the importance of Chasles, we may safely assume that the memoir was read by some of the experts on mechanics who introduced

---

[14] The theorem is attributed to Chasles, for example, in Webster [1942: 213]. According to Loria, it had been previously found by Gergonne, Möbius and Giorgini [Loria 1893: 30]. Chasles himself remarks that it had been enunciated in a slightly different form by Poinsot in his *Statique* [Chasles 1830: 102]. Probably the first general formulation, with three different proofs, was given by Bordoni [1821: 115].

However, taking into account the importance of Chasles, we may safely assume that the memoir was read by some of the experts on mechanics who introduced ideas related to vector calculus in their textbooks at the end of century.

Acknowledgment

I am indebted to Massimo Corradi, Federico Foce, Antonio Becchi and the Associazione Edoardo Benvenuto for having invited me to contribute to the present book. My warmest thanks go to Kim Williams for her linguistic advice. I also want to express my gratitude to Prof. Livia Giacardi, of the University of Turin, for her worthwhile suggestions and her help in reading the proof sheets. The research for this paper was carried out as part of the Turin unit of the Progetto nazionale MIUR "Storia delle scienze matematiche".

Bibliography

BINET, JACQUES PHILIPPE MARIE. 1815. Mémoire sur la composition des forces et sur la composition des moments. *Journal de l'École Polytechnique*, vol. X, no. XVII: 321-348.

BORDONI, ANTONIO. 1821. Sui sistemi di due forze equivalenti fra loro e ad un qualsivoglia. *Giornale di Fisica, Chimica, Storia Naturale, Medicina ed Arti di Pavia*, vol. IV: 114- 132.

CAPARRINI, SANDRO. 2002. The Discovery of the Vector Representation of Moments and Angular Velocity. *Archive for History of Exact Sciences*, vol. 56: 151-181.

CARNOT, LAZARE NICOLAS MARGUERITE. 1803. *Géométrie de position*. Paris: Duprat.

CHASLES, MICHEL. 1830. Mémoire de géométrie pure, sur les systémes de forces, et les systèmes d'aires planes; et sur les polygones, les polyèdres, et les centres des moyennes distances. *Correspondance mathématique et physique*, vol. VI: 92-120.

―――. 1837. *Aperçu historique sur l'origine et le développement des méthodes en géométrie*. Bruxelles: M. Hayez. Repr. Paris: Gauthier-Villars, 1875 and 1889. Repr. Paris: Gabay, 1989. (German translation: *Geschichte der Geometrie; hauptsächlich mit Bezug auf die neueren Methoden*. Halle: Gebauer, 1839. Repr. Wiesbaden: Sändig, 1968. Russian translation: *Istoriceskij obzor proischozdenija i razvitija geometriceskich metodov*. Moscow: Mamontov 1871-72.)

―――. 1847. Théorèmes généraux sur les systèmes de forces et leurs moments. *Journal de mathématiques pures et appliquées*, vol. XII: 213-224.

―――. 1870. *Rapport sur les progrès de la géométrie*. Paris: Imprimerie Nationale.

CROWE, MICHAEL JOHN. 1967. *A history of vector analysis: the evolution of the idea of a vectorial system.* Notre Dame, Indiana: University of Notre Dame Press. Repr. New York: Dover, 1985, 1994.

FRANÇAIS, FRANÇOIS JOSEPH. 1808. Mémoire sur la transformation des coordonnées. *Journal de l'École Polytechnique*, vol. VII, no. XIV: 182-190.

GIORGINI, GAETANO. 1813. Questions de mathématiques et de physique proposées au concours général des lycées de Paris, année 1812. *Correspondance de l'École Polytechnique*, n. 5, vol. 2: 439-445.

———. 1814. Démonstration de quelques théorèmes de géométrie. *Correspondance de l'École Polytechnique*, vol. 3, n. 1: 6-9.

———. 1820. *Teoria analitica delle proiezioni.* Lucca: Tipografia Ducale di Francesco Bertini. Repr. *Atti dell'Accademia Lucchese di Scienze Lettere ed Arti*, vol. I (1821): 29-96.

———. 1828. Sopra alcune proprietà de'piani de' momenti principali e delle coppie di forze equivalenti. *Memorie di matematica e di fisica della Società Italiana delle Scienze*, vol. XX: 243-254.

HACHETTE, JEAN NICOLAS PIERRE. 1809. Sur la transformation des coordonnées. *Correspondance sur l'École Polytechnique*, vol. 2: 6-13.

———. 1811. Note sur la transformation des coordonnées obliques en d'autres obliques ayant la même origines. *Correspondance sur l'École Polytechnique*, vol. 2: 247-249.

———. 1813. *Traité des Surfaces du second degré.* Paris: Klostermann.

———. 1817. *Éléments de Géométrie à trios dimensions.* Paris: L'auteur.

LORIA, GINO. *1893.* Intorno la vita e le opere di Gaetano Giorgini. *Giornale di Matematiche*, vol. 31: 23-30. Repr. in *Scritti, conferenze, discorsi sulla storia delle matematiche,* Padua: Cedam, 1936.

PAULI, WOLFGANG. 1958. *Theory of relativity.* Translated from the German by G. Field, with supplementary notes by the author. New York: Pergamon Press. Repr. New York: Dover, 1981.

POISSON, SIMEON DENIS. 1811. *Traité de Mécanique.* 2 vols. Paris: Courcier.

REICH, KARIN. 1996. The Emergence of Vector Calculus in Physics: the Early Decades. Pp. 197-210 in *Hermann Günther Graßmann: (1809-1877); visionary mathematician, scientist and neohumanist scholar; papers from a sesquicentennial conference.* Gert Schubring, ed. Dordrecht: Kluwer Acad. Publ.

WEBSTER, ARTHUR GORDON. 1942. *The dynamics of particles and of rigid, elastic, and fluid bodies; being lectures on mathematical physics.* 3rd edition. New York: G. E. Stechert & Co.

THE ANCIENTS' INFERNO:
THE SLOW AND TORTUOUS DEVELOPMENT OF
'NEWTONIAN' PRINCIPLES OF MOTION IN THE EIGHTEENTH
CENTURY

Giulio Maltese[1]

In his *Introduction to the History of Structural Mechanics* Edoardo Benvenuto provided us with the suggestive image of a science growing according to two laws. The first one is represented by the *Apollo's arrow*: "The arrow hits the quarry and kills it". This is a metaphor of science seen as an ever-growing machinery whose ultimate goal is to solve particular questions and make progress possible. The second law is represented by the *Ancients' Inferno*, "where Danaides, Ixions and Sysiphes eternally labour, filling bottomless hogsheads and lifting weights which always fall again: in its theoretical development throughout the centuries, scientific work seems to be concerned with defining again and again a small list of basic words which give form to the foundations of theories". Such a twofold vision of science applies surprisingly well to the way the so-called "Newtonian" mechanics is today seen and taught and the way, in the eighteenth century, it grew out of the mist of confusion, detours, false starts, and repetitive definitions of "a small list of basic words", of which "force" is probably the best example. The painstaking task of ancient *mécaniciens*, and in particular of those forming the so-called "Basel school" has been told by Clifford A. Truesdell in his works, and especially in *The Rational Mechanics of Flexible or Elastic Bodies 1638-1788* and in his celebrated *Essays in the History of Mechanics*. It is my purpose to recall, in Benvenuto's and Truesdell's own words, the complex and tortuous development of ideas eventually leading to the modern laws of 'Newtonian' mechanics as opposed to the view of those laws – and of the additional concepts they imply, such as "force", or "relativity", or "reference system" – we tend to have today, i.e., of a part of a mathematical machinery laid down at some point in time and since then restless acting as *Apollo's arrow*.

Introduction

It's a great honour for me to pay homage to two *savants* such as Professors Truesdell and Benvenuto. They both played a major role in providing directions and opportunities for my activity concerning the history of mechanics. I will therefore begin this paper with a personal recollection.

Ever since I was an undergraduate student in physics, I wondered how the various principles of mechanics came to be discovered and, even more puzzling, why so many different "versions" of those principles were needed in the development of mechanics. To be sure, the issue of constraints was a clear one, and mechanical principles used to be classified according to how they took into

---

[1] Group for the History of Physics, Department of Physics, University "La Sapienza," Rome, ITALY

account the forces originating from constraints. However, since the way of teaching mechanics was strictly non-historical, I was left with my curiosity unsatisfied as to how and why the *mécaniciens* passed from Newton's to D'Alembert's principles (I had no doubt that Newton's laws of motion were created by Newton himself) and eventually to Lagrange's, to Hamilton's and to Hamilton-Jacobi's principles. Such a vague inclination towards the history of mechanics remained in the background of my mind till I discovered Truesdell's works, and in particular his *Essays in the History of Mechanics* and later *The Rational Mechanics of Flexible or Elastic Bodies*. There I discovered that not only the path to the modern principles of mechanics was long and tortuous, with frequent false starts and detours, but also that the celebrated "Newtonian mechanics" was completed almost a century after the publication of Newton's *Principia*, thanks to the work of the so-called "Basel school," formed by Leonhard Euler and by the Bernoullis. I learned that Truesdell called for a "Program of Rediscovery of the Rational Mechanics of the Age of Reason" and that, as he said,

> others [than Truesdell himself] might be led to share the pleasure of discovery for themselves [...] if now I give out this short survey of the riches waiting for anyone willing to take down the dusty volumes and read them, as they have not been read for two centuries, with consequences of the mathematics [Truesdell 1968:87].

I actually became involved in such a program of rediscovery, and I took great pleasure in reading those volumes (some of them were actually very dusty) that Truesdell had mentioned. I focused on the history of the Newtonian principles of motion in the eighteenth century, and in particular on the so-called "Newton's second law of motion."

Before meeting Edoardo Benvenuto I became familiar with his splendid book, *Introduction to the History of Structural Mechanics*, that he sent me after receiving of mine [Maltese 1992a]. We met some time later, and he proposed that I teach some history of dynamics to his students, both undergraduate and graduate. Even if his principal interest in history of science concerned history of statics and of structural mechanics, he was fascinated by the complexity of the history of dynamics, devoting attention and time to listen to my reports on the status of my work.

I soon noticed that a number of ideas explained in the initial chapters of his *Introduction to the History of Structural Mechanics* might well apply to the history of dynamics as well, and we kept discussing the ambiguity of the concept of force or on the complexity of the development of the principles of motion in the seventeenth and in the eighteenth centuries. This is why, when I was invited by the organizers of this conference to give a talk, I chose as my subject the history of the principles of the so-called "Newtonian" laws of motion in the eighteenth century. In so doing, I will have the opportunity to quote some of the most important of Truesdell's and Benvenuto's ideas on the development of mechanics and of science in general.

Two laws describing the growth of science

An adequate starting point is Benvenuto's remark about "an interpretative key of history which today dominates the 'philosophy of science.'" According to this,

> *science grows according to two laws. The first law is represented by the image of Apollo's arrow. The arrow hits the quarry and kills it. In its practical development, science succeeds in solving a lot of particular questions and therefore these questions disappear as such. The second law is represented [by the so-called] land of Ancients' Inferno, where Danaides, Ixions and Sysiphes eternally labour, filling bottomless hogsheads and lifting weights which always fall again: in its theoretical development throughout the centuries, scientific work seems to be concerned with defining again and again a small list of basic words which give form to the foundations of theories* [Benvenuto 1991:9].

Benvenuto continues,

> *A two-fold task is [therefore] assigned to a general 'history of science': on one side, to record in chronological order the experimental discoveries and the solved theoretical questions which provide modern science and technology with answers useful to man's needs; on the other side, to demonstrate that these particular achievements have not put an end to the work of Sisyphes and Danaides, but made their work more and more tiring, burdening the same basic vocabulary of science with ever new claims and aims* [Benvenuto 1991:9].

One may well wonder, as Benvenuto does, "what connection there may be between the *real* history of mechanics and this philosophical representation (especially the one denominated '*the land of Ancients' Inferno*')" [Benvenuto 1991:9].

A suitable answer to this question may well come if one looks at how mechanics is taught today and how mechanics actually developed. On the one hand, mechanics is frequently presented as a sort of mathematical machinery laid down at some [definite] point in time and since then restlessly acting as *Apollo's arrow*. On the other hand, the work of many historians of science has provided us with convincing evidence that mechanics grew out of the mist of confusion, detours, false starts, and repetitive definitions of "a small list of basic words" (of which "force" is probably the best example) and therefore its history can be best represented by *the land of Ancients' Inferno.*

The *Apollo's arrow* applies to the way mechanics is seen, while *the land of Ancients' Inferno* applies to the history of mechanics. These two views are simply two sides of the same coin, and neither is sufficient *per se* to give a reasonable representation of the science of motion. It follows that the art of teaching mechanics cannot be complete without some historical presentation, due to the distance between the two views of mechanics.

A case in point is represented by the so-called "Newtonian" mechanics, i.e., that tradition of mechanics resting upon the principles of linear and angular momentum: where did it come from? Was it a creation of Newton alone? Was it created in a short period of time?

The modern view may well be summarized by the following quotation: "today the textbooks of physics [...] present the laws of motion in almost exactly the same form as was used by Newton in the *Principia.*" [Eisenbud 1958:144]. My paper will be devoted to analysing that "almost exactly."

Newton's second law of motion

Let's see what Newton actually wrote in his *Principia.* At the beginning of Book I stand the famous Axioms, or Laws of Motion. The second law reads, "the change of motion is proportional to the motive force impressed, and it takes place along the right line in which that force is impressed" [Newton 1687:114]. In modern terms, this definition corresponds to $F=\Delta(mv)$, where *motus* (momentum) is mv. However, this definition is far from univocal, for Newton uses a different definition of "change of motion" (*mutatio motus*) according to the kind of motive force under consideration. He gives three examples of motive force: pressure, collision and centripetal. After that, he devotes himself to the examination of centripetal force, for which he gives the following definition: "the motive quantity of centripetal force is a measure of it and is proportional to the motion that it generates in a given time" [Newton 1687:99]. Thus, for the centripetal force the second law of motion seems to hold in the form $F=ma$.

To summarize: in the *Principia* the second law of motion is used in one of the two forms:

$$(1) \quad \begin{aligned} F &= \Delta\,(mv) \\ F &= mdv\,/\,dt = ma \end{aligned}$$

The reason for such a dichotomy stems from the two ways force is conceptualized, i.e., either as an entity acting discontinuously, as it happens for two colliding bodies, or continuously, as the force of gravity acting on a falling bodies. Newton accomplished a major conceptual leap passing from the concept of force of a body in motion to the concept of force that changes the state of motion of a body. In so doing, however, he could not escape one of the main ambiguities affecting mechanics in the seventeenth century. Nevertheless, Newton succeeded in building the splendid edifice of his mechanics. It is to be remarked, anyway, that the modern science of motion could not have been grounded on the second law of motion unless the ambiguity in its formulation had been removed.

Another question one may ask concerns the kind of mechanical systems that are dealt with in the *Principia*. Or, in other words: how complete is Newton's masterpiece from the point of view of the variety of mechanical systems? According to Truesdell,

> In Newton's Principia *occur no equations of motion for systems of more than two free mass-points or more than one constrained mass-point; [...] the spinning top, the bent spring, lie altogether outside Newton's range. [...] Newton gives no evidence of being able to set up differential equations of motion for mechanical systems* [Truesdell 1968: 92-93; Truesdell's emphasis].

Finally, is it important to stress the difference between the modern form of the *momentum principle* and the second law of motion in the *Principia*? The answer comes from Cannon and Dostrovsky:

> To grasp how it was that 'Newton's Second Law' had a different appearance before the middle of the eighteenth century, one should distinguish between systems in one and in many degrees of freedom [...] 'Newton's Second Law,' as it was understood before the middle of the eighteenth century, [was] a condition on a particular motion of a mechanical system: the mass of an element of the system multiplied by its acceleration must be equal to the force which that element experiences during a given motion of the entire system. This is a condition on a given motion and it is not a system of dynamical equations. In practice, [Newton's Second Law] is a consistency condition that typically leads to a constant of a particular motion such as the period of an oscillation* [Cannon and Dostrovsky 1981:1-2].

We therefore see that Newton's laws of motion in his *Principia* are far from those generally accepted today as the "laws of Newtonian mechanics". There is a big discrepancy between what modern readers (and teachers) deem Newtonian and what Newton actually did. Where did it come from? A reasonable answer comes from Truesdell, who, in his *Essays*, put the genesis of the modern attitude towards Newton's work in Mach's celebrated book *Die Mechanik in Ihrer Entwicklung, historisch-kritisch dargestellt* first published in 1883. In his famous essay, which, to be sure, Mach himself did not intend as a mere historiographical reconstruction of the development of mechanics, he asserted that,

> *Newton discovered universal gravitation and completed the formal enunciation of the mechanical principles now generally accepted. Since his time no essentially new principle has been stated. All that has been accomplished in mechanics since his day has been a deductive, formal, and mathematical development of mechanics on the basis of Newton's laws.*

And he added,

> *The principles of Newton suffice by themselves, without the introduction of any new laws, to explore thoroughly every mechanical phenomenon practically occurring* [Mach 1968:106 and 291; translated in Truesdell 1968:86].

According to Truesdell, the seed of Mach's interpretation comes from Lagrange's *Méchanique Analitique*, published in 1788, just a century after the *Principia*:

> *Every popular history glows in generality over it and quotes Hamilton's judgment that it is 'a kind of scientific poem.' In two ways this book has cut off from earlier researches. First, the working scientist has accepted the* Méchanique Analitique *[...] as the final repository of all the mechanics that went before it and has not felt need to look behind it to its sources or around it to discover what it left out. Second, Lagrange included in the* Méchanique Analitique *short histories of statics, dynamics, and fluid mechanics, and these, too, have been accepted as final in outline if not in detail. Nearly all of Mach's pages [at least in the first edition of his* Mechanik*] about work done before 1800 seem to have been filled either by use of the sources cited by Lagrange or by plausible conjecture how discoveries of that kind must surely have been made, or at least would be made by a right-thinking person. [...] Judged by their works, the historians seem to think mechanics stopped with Newton, except for the formal developments which are to be found in textbooks and are there associated with later names. No great historical effort has been spent upon the growth of mechanics since Newton's time.*

The scientists, in so far as they take any note of history at all, not only have shared the historians' neglect of the later mathematical development of mechanics but also, in the main, have ignored what historians have learned about the earlier periods and have rested content with Mach's whole view or a rudimentary abstract of it. Between Lagrange and Mach, between the historians and the scientists, the Age of Reason is left in the Dark Ages of the history of mechanics. Yet, it is [not] the primitive mechanics of Newton [...] that we are taught as the most successful, the most thoroughly proved and understood, and the most perfect of the sciences of nature – the prototype and paradigm of a mathematical theory for physical phenomena. Rather, it is the easier parts of the rational mechanics of the Bernoullis, Euler, and their successors. Whence did it come? Why was it sought? How was it made? Where did it succeed, where did it fail? [...] To answer these questions, I call for a program of rediscovery of the Rational Mechanics of the Age of Reason [Truesdell 1968:86-87].

Ambiguities, obstacles, and advances on the path towards rational mechanics up to 1750

At the beginning of the eighteenth century there were several obstacles to the progress of the science of motion. Some of them were not altogether independent of each other. For example, we have seen that the ambiguity in the form of the second law of motion stemmed from the apparent non-unique way of representing force as an entity acting either continuously or discontinuously. Another ambiguity about the nature of matter was strictly related to the previous one: are bodies perfectly rigid or are they continuously deformable? If bodies are deformable, what are the differences in behaviour between elastic and anelastic bodies? Clearly, if one believes in the existence of perfectly hard bodies, one will favor the first of equations (1) as the correct form of the second law of motion [See Scott 1959 and Maltese 1992a: chapter 6].

Additionally, much needed advances concerned the representation of motion and of physical laws. On the one hand, 'intrinsic' coordinates (taken along directions tangent and normal to the curve described by a moving bodies) were still in use in the 1730s, and they were much less useful, elegant and powerful than Cartesian rectangular coordinates. On the other hand, at the beginning of the eighteenth century, physical quantities were often represented by segments and physical laws by proportions: the algebraic notation was still to come [See Ravetz 1961 and Maltese 1992a: 79 and 137-138].

The toughest obstacle towards the modern science of motion was probably the lack of a way to deal with mechanics of extended bodies. Speaking of mechanics of continua, Truesdell has recognized a potential difference in the power of mathematical methods available:

> It is instantly plain that <u>the language of our subject is partial differential equations</u>. Now the English with their fluxions and the Continentals with their differentials were, in principle, on a par so long as problems involved but one independent variable. But no problem of the dynamics of continua [...] is of this kind. The <u>idea</u> of a partial derivative was indeed known [...] Lacking was a formal <u>calculus of partial derivatives</u>, and for developing such a calculus the fluxional concepts, while indeed admissible, were not conducive. What was needed was a man who could <u>express and master the Newtonian view of mechanics in Leibnizian partial differentials</u>. This man was Euler [Truesdell 1960:141; Truesdell's emphasis].

In his *Introduction to the History of Structural Mechanics* Benvenuto put forward the interesting idea that,

> the entire history of mechanics up to the close of the eighteenth century can probably be told in terms of particular objects which embody a concept – which give a physical image to a line of thought and testify to a principle. Early theoretical thinking about statics and mechanics took as its references particular objects, things like the lever, used since ancient times as necessary tools [Benvenuto 1991:4].

This idea is also extraordinarily well suited for the development of dynamics in the eighteenth century. Let us think for a moment to the so-called 'dynamicized' balance and its importance in the history of the law of angular momentum, or to objects like the compound pendulum, which through investigations on the determination of the 'center of oscillation' led to one of the first successful researches on the dynamics of extended bodies. Just a partial list is formed by other objects: the vibrating string, the oscillating chain, the motion of mass-points into rotating tubes (a mechanical system widely studied in the 1740s), the rigid body and the elastic line. We shall soon meet again most of these objects. I will now outline some of the most important advances in the science of motion in the initial decades of the eighteenth century, that paved the way to the appearance of the 'Newtonian' laws of motion.

Leonhard Euler was in 1730 the first to give a substantial contribution by treating the laws of collision – a typical problem earlier dominated by the version of the second law of motion that can be written $F=\Delta(mv)$ – according to the law

F=mdv/dt, by resorting to the *principle of continuity* according to which *natura non facit saltus* and perfectly hard bodies do not exist, since every body is deformable when subject to a force [Euler 1730].

In 1736 the first of great Euler's treatises on mechanics, entitled *Mechanica, sive motus scientia analytice exposita*, was published. Truesdell remarked that,

> *The* Mechanics *made three additions to the principles. First, while Newton had used the word 'body' vaguely and in at least three different meanings, Euler realized that the statements of Newton are generally correct only when applied to masses concentrated at isolated points; he introduced the precise concept of mass-point, and this is the first treatise devoted expressly and exclusively to it. Second, he is the first to recognize explicitly the acceleration and to study it as a kinematical quantity defined in a motion along any curve. Third, he employs the concept of vector or 'geometrical quantity,' a directed magnitude, as pertaining not only to static force, for which it was familiar, but also to velocity, acceleration and other quantities* [Truesdell 1968:107].

However Euler was well aware that the principles he had – basically the second law of motion which he generally used in the form *dv=adt* – was adequate to describe only the motion of mass-points:

> *Those laws of motion which a body observes when left to itself in continuing rest or motion pertain properly to infinitely small bodies, which can be considered as points. Indeed, in bodies of finite magnitude, whose several parts are endowed with various motions, a given part will try to observe these laws, which however on account of the state of the body is not always possible. Therefore the body will follow that motion which is composed from the endeavors of the several parts, and from insufficiency of the principles this cannot yet be determined, but its treatment is to be left to the sequels* [Euler 1736, vol. I: 38-39; translated in Truesdell 1955:IX].

An important vein of researches in mechanics is represented by the works on the 'center of oscillation'. In 1673 Huygens was the first to solve the problem of finding the length of the simple pendulum isochronous to an extended oscillating body. In 1703 Jacob Bernoulli gave a solution of this problem in an important paper [Bernoulli, Jacob 1703], titled *Demonstratio centri oscillationis ex natura vectis*, which, according to Truesdell, was

> *second only to the Principia itself in influence on the later growth of the discipline [...] Three major ideas – Truesdell continues –*

are foreshadowed in Jacob Bernoulli's method: 1. to determine the motion of a constrained system, introduce forces which maintain the constraints; 2. the accelerations of bodies if reversed in sign are equivalent to static forces (per unit mass); 3. not only equilibrium of forces but also equilibrium of moments is necessary in dynamic as well as static problems [Truesdell 1968:104].

Between 1713 and 1728 Johann Bernoulli and Brook Taylor published their researches on the vibrating string [Taylor 1713 and Bernoulli, Johann 1728]. Their aim was to find the center of oscillation of the curve formed by the fundamental mode of a vibrating string. Because a vibrating string is a flexible body, they had first to find its shape. That was the context of the first researches on this very important problem [Maltese 1992a, 1992b]. While the integration of the differential equation for the motion of a vibrating string presented no major problems, this was not the case for an oscillating chain, a problem tackled in 1733-1734 by Daniel Bernoulli [Bernoulli, Daniel 1733 and 1735]. He decided to adopt a series integration, and this led him to the analytical discovery of the higher modes of vibration [Maltese 1992a:chapter 4]. According to Truesdell,

> *these two papers are as fine as any Daniel Bernoulli ever wrote, and they bring a magnificent contribution to the theory of vibrations [...] It is instructive to follow the difficult and clumsy steps by which Bernoulli demonstrated his results. In addition to being as great an expert on mechanics as any then living, he was an especially thorough student and admirer of Newton's* Principia. *Those who parrot the conventional view that Newton's principles suffice to solve all problems of mechanics should read these papers, from which it is most plain that if such be the case, Daniel Bernoulli, at least, did not know it in 1733* [Truesdell 1960:141].

In 1737 Euler conceived a major treatise on the science of naval construction. By 1741 he had completed his work. However, he was able to publish it only in 1749, when maybe the most important results as far as naval science is concerned had already been published by Euler's rival in this field, Pierre Bouguer. However, Euler's *Scientia Navalis* contains some initial and very important results in the mechanics of rigid bodies. It was during his work on this treatise that Euler realised that a rigid body can accomplish two motions: a translatory motion of its center of mass and a rotatory motion about an axis passing through its center of mass. These motions are independent, and in order to study the rotatory one Euler devised to work in the frame of reference where the center of mass is at rest, since no force applied to the center of mass can perturb the rotatory motion:

Motus autem, quem potentiae centro gravitatis inducunt, non est solus effectus quem potentiae in corpore producunt, neque ad effectum potentiarum corpori applicatarum sufficit motum centri gravitatis nosse. Interea enim dum centrum gravitatis vel quiescit, vel descripto modo movetur, fieri potest, ut reliquae corporis partes circa centrum gravitatis revolvantur, motumque gyratorium accipiant, quippe quo motu centri gravitatis motus non turbatur [Euler 1749:141].

This important discovery led Euler to assert that the two motions had to be described by two independent principles, of which the one describing the rotatory motion was to be built analogously to the principle describing the linear motion:

Cognita autem vi gyratoria, si ea per elementum temporis multiplicetur, productum dabit elementum celeritatis gyratoriae hoc tempusculo genitum; unde [...] motus gyratorius eodem modo determinari poterit quo motus corporis a potentia sollicitati in directum [Euler 1749:75].

The analogy between the two principles is clearly emphasized by Euler in a (second-)prize-winning essay, published soon after the completion of his *Scientia Navalis*, in which he used extensively the newly discovered principle of angular momentum:

Le premier de ces momens donc, qui est celui des puissances sollicitantes, divisé par le moment de la matiere, donnera la force de rotation, de la même maniere que dans les mouvemens progressifs la puissance même divisée par la matiere même du corps, exprime la force accélératrice. Cette grande analogie mérite bien d'etre remarquée [Euler 1741:51].

In addition to the above-mentioned works, Euler adopted his newly-discovered principle of angular momentum in other important papers, like those on the motion of mass-points in rotating tubes [see for example Euler 1746 and Maltese 2001: chapter 5.3] and the work he presented in 1737 on the *percussio excentrica*, where he asserted,

I attained my goal [namely, to describe the motion of two two-dimensional extended colliding bodies, moving on a plane] by making use of new principles of mechanics that I recently discovered [Euler 1737:8].

According to the celebrated, albeit sometimes inaccurate, short histories of statics, dynamics and fluid mechanics that Lagrange placed at the beginning of

his *Mechanique Analitique*, it was MacLaurin who first expressed the second law
of motion employing space-fixed Cartesian coordinates in his 1742 *Traité des
fluxions* [Lagrange 1888:243]. Truesdell, however, questioned the correctness of
Lagrange's statement, claiming that "in the book of MacLaurin [...] there is
neither any general statement of the laws of mechanics nor any example
formulated in Cartesian co-ordinates" [Truesdell 1960:252]. In Truesdell's view,
the first to use Cartesian rectangular coordinates to solve a mechanical problem
was Johann Bernoulli in 1742; but this can safely be antedated to 1735 [See
Maltese 1992a:253-254, Maltese 2000, and Maltese 2001:96-97]. In 1747 Euler
was the first to express the second law of motion in rectangular Cartesian
coordinates (i.e. with three equations along three fixed axes) [Euler 1747].
Truesdell has emphasized the importance of the transition to Cartesian
rectangular coordinates:

> *the importance of the use of Cartesian co-ordinates lies deeper than
> in mere simplicity; in these co-ordinates the addition of vectors
> located at different points is so natural as to become customary at
> once, and the possibility of performing this addition lies at the heart
> of the classical conception of space-time; once the laws of
> mechanics is stated [in rectangular Cartesian co-ordinates],
> however, discovery of the properties of the total momentum,
> moment of momentum, and kinetic energy for any system of mass-
> points becomes trivial, as any beginner knows [...] today this
> possibility [the use of Cartesian co-ordinates] is so obvious that
> many scientists seem to believe that Newton himself used Cartesian
> co-ordinates, but of course this is not so* [Truesdell 1960:252-253].

According to David Speiser, "The essence of Newton's great discovery is that
forces are vectors [...] Cartesian coordinates are the closest surrogate for an
abstract vector calculus" [Speiser 1983:38-39].

In his *Hydraulics*, written probably in 1740, the elderly Johann Bernoulli
made another step forward isolating an element of fluid and applying to it the
principle of linear momentum, thereby obtaining a differential equation that,
integrated, gives the solution [Bernoulli, Johann 1742]. Bernoulli was the first to
apply Newton's second law of motion to an infinitesimal element of a
continuum. Truesdell remarked that,

> *it is easy enough nowadays to claim that Newton gave the necessary
> principles and to call all of the classical mechanics 'Newtonian', but
> neither Newton, nor any of his disciples or rivals was able to give a
> satisfactory treatment of any problem involving the dynamics of
> deformable masses. It was John Bernoulli [...] who, in 1740, first
> employed the method of calculating the force acting on an*

infinitesimal element. This it was which Euler recognized as 'the true and genuine method' and proceeded to apply at once [Truesdell 1955:XLI].

During years 1747-1749 a substantial effort was devoted by Euler, Clairaut, and D'Alembert to the problem of the motion of lunar apogee. All of them initially questioned the validity of Newton's law of universal gravitation, and all of them later found errors in their calculations. In their researches, however, the existing principles of motion proved to be inadequate. First of all, the transition to Cartesian rectangular coordinates proved to be necessary. Secondly, when studying the motion of celestial bodies, what was necessary was a relationship between position and force, rather than a relationship force-velocity, as was the case employing the relationship *dv=adt*. Thirdly, existing principles, including the law of angular momentum found by Euler in his *Scientia Navalis,* were inadequate to determine the motion of celestial bodies having a non-spherical shape or inhomogeneous mass distributions. In other words, when the rotatory motion of a rigid body occurs about a non-fixed axis, Euler's method were inadequate. Euler was aware of this status of affairs since the time of his *Scientia Navalis.* In 1749, however, he drew substantial inspiration from D'Alembert's *Recherches sur la Précession des Equinoxes et sur la Nutation de l'Axe de la Terre dans le Système Newtonien* [D'Alembert 1749]. D'Alembert obtained the total moment of elementary forces acting on a body with respect to a given axis by means of a clever procedure of summation of the contributions coming from each mass element of the body. Euler stated that D'Alembert had used "lofty rules" (*multo sublimiores regulae*) and soon set out to use the heart of D'Alembert's method [See Wilson 1987, Maltese 2001: chapter 7].

Euler's *Découverte d'un nouveau principe de Mécanique* and the first equations of motion for a rigid body

On 3 September 1750 Euler presented to the Berlin Academy of Sciences a paper entitled *Découverte d'un nouveau principe de Mécanique* [Euler 1750], where he set down the momentum principle written in Cartesian rectangular coordinates

$$2mddx = \pm P \, dt^2$$
$$(2) \quad 2mddy = \pm Q \, dt^2$$
$$2mddz = \pm R \, dt^2$$

as the axiom which "includes all principles of mechanics." Truesdell, who was the first to call attention to this fundamental paper, remarked:

We may justly wonder that it took more than sixty years for so simple an extension of Newton's ideas, but the literature of mechanics does not permit us to doubt that it did. As often happens in the history of science, the simple ideas are the hardest to achieve: simplicity does not come of itself but must be created [Truesdell 1960:251].

In Equations (2) mass m can be finite or infinitesimal; Euler had finally realised that Equations (2) can be applied to every particle of every body.

Since Euler claims that Equations (2) "include all principles of mechanics", one may well wonder what happened to the law of angular momentum, expounded by Euler on a par with respect to the principle of linear momentum in his *Scientia Navalis*. The answer comes in the second part of the 1750 paper, where Euler gives for the first time his method to obtain the equations of motion for a rigid body rotating about a non-fixed axis. First of all, one has to start from the fact that Equations (2) "include all principles of mechanics" insofar as every body can be considered as composed of small corpuscles whose motion can only be rectilinear and can be therefore described by Equations (2). Secondly, for extended bodies one is confronted with the task of connecting "microscopic" to "macroscopic" motion and the action of "elementary" forces to the action of "external" forces. By "elementary forces" Euler means the quantity $dmddx/dt^2$. The sum of motions due to elementary forces is the "microscopic motion." On the other hand, the external forces acting on an extended body determine its "macroscopic motion." The two sets of forces are equivalent, since they produce exactly the same motion. Thirdly, let's take, for example, the case of rigid bodies. Let F_x and F_x^e be the components along the x-axis of the resultant of elementary and external forces, respectively acting on a rigid body. Let also N_x and N_x^e be the components along the x-axis of the resultant of the moment of elementary and external forces, respectively. Then, Euler's principle reads:

$$
(3) \qquad
\begin{aligned}
F_x^e &= F_x \\
N_x^e &= N_x
\end{aligned}
$$

Euler explicitly states that the resultant of the forces due to constraints has to be zero, in the case of the rigid body. He also implicitly assumes that the same holds for the moments of those forces. Then, using the second law of motion (2) in Eqs. (3) Euler obtains:

$$
(4) \qquad
\begin{aligned}
F_x^e &= F_x = \int dF_x = \int 2dm \frac{ddx}{dt^2} \\
N_x^e &= N_x = \int dN_x = \int ydF_z - zdF_y = \int 2ydm \frac{ddz}{dt^2} - \int 2zdm \frac{ddy}{dt^2}
\end{aligned}
$$

To complete Euler's procedure it now suffices to substitute in Equations (4) for second differentials of the coordinates the expressions that can be derived describing the kinematics of the motion of the rigid body:

$$ddx = (zd\omega_y - yd\omega_z)dt + [\omega_z\omega_x z + \omega_y\omega_x y - (\omega_z^2 + \omega_y^2)x]\,dt^2$$

(5) $$ddy = (xd\omega_z - zd\omega_x)dt + \left[\omega_y\omega_x x + \omega_z\omega_y z - (\omega_x^2 + \omega_z^2)y\right]dt^2$$

$$ddz = (yd\omega_x - xd\omega_y)dt + [\omega_z\omega_x x + \omega_z\omega_y y - (\omega_y^2 + \omega_x^2)z]\,dt^2$$

Substituting Equations (5) into (4) allows Euler to get his first equations for the motion of a rigid body:

(6)

$$\frac{N_x}{2} = I_{xx}\frac{d\omega_x}{dt} - I_{xz}\frac{d\omega_z}{dt} - I_{xy}\frac{d\omega_y}{dt} - \omega_x\omega_y I_{xz} + \omega_x\omega_z I_{xy} + I_{yz}(\omega_z^2 - \omega_y^2) - \\ -\omega_y\omega_z(I_{yy}^2 - I_{zz}^2)$$

$$\frac{N_y}{2} = I_{yy}\frac{d\omega_y}{dt} - I_{xy}\frac{d\omega_y}{dt} - I_{yz}\frac{d\omega_x}{dt} - \omega_x\omega_y I_{xy} + \omega_y\omega_z I_{yz} - I_{xz}(\omega_z^2 - \omega_x^2) - \\ -\omega_x\omega_z(I_{zz}^2 - I_{xx}^2)$$

$$\frac{N_z}{2} = I_{zz}\frac{d\omega_z}{dt} - I_{xz}\frac{d\omega_x}{dt} - I_{xy}\frac{d\omega_y}{dt} - \omega_y\omega_z I_{xz} - \omega_x\omega_z I_{xy} - I_{yz}(\omega_x^2 - \omega_y^2) + \\ +\omega_x\omega_y(I_{yy}^2 - I_{zz}^2)$$

So he can conclude that,

> *Ce seront donc ces trois formules, qui contiennent les nouveaux principes de Mécanique, dont on a besoin pour déterminer le mouvement des corps solides, lorsque l'axe de rotation, autour duquel ils tournent, ne demeure pas immobile* [Euler 1750: 105].

The final steps to the cardinal equations of mechanics

Euler's *Découverte d'un nouveau principe de Mécanique* marks a break in the history of mechanics. As far as principles of motion are concerned, Euler's fourteen-year battle against the motion of a rigid body was over. What was still wanting was the representation of motion, as Equations (6) were by far too complicated. Furthermore, as we will see, a better representation of motion would have permitted Euler, some 25 years later, to separate principles of motion from its representation.

In his 1750 paper Euler had used space-fixed axes. Only one year later, he presented a paper where he employed body-fixed axes. Furthermore, he was aware of the existence of one axis of free rotation (i.e., such that if a rigid body rotates around it, it will indefinitely rotate around it, as no force is generated that induces any change in the direction of the axis of rotation) [Euler 1751]. In 1755 Andres Segner discovered the existence of *three* axes of free rotation. This result induced Euler to study the motion of a rigid body in greater detail. In 1758 he came up with two important papers on the motion of a rigid body. The first, titled "Recherches sur la connoissance mécanique des corps" contained not only the analytical demonstration of the existence of three axes of free rotation, but also a sort of Eulerian *manifesto* concerning the separation of knowledge of bodies. Euler classified the knowledge about bodies in "geometrical" (concerning the extension and the figure of a body), "mechanical" (concerning the inertial properties of bodies) and "physical" (concerning the qualities other than inertia of the matter forming bodies). What has to be emphasized is that Euler considered the mechanical knowledge of a class of bodies on a par with respect to the knowledge of the principles of motion. After this introduction to the description of the inertial properties of rigid bodies, in his second paper of 1758 Euler gave for the first time the equations of motion for rigid bodies that were to be named after him, using a reference system whose axes coincided with the axes of inertia of the body under examination:

$$I_{xx}\frac{d\omega_x}{dt} + \omega_y\omega_z(I_{zz}-I_{yy}) = 2gN_x$$

$$(7) \qquad I_{yy}\frac{d\omega_y}{dt} + \omega_x\omega_z(I_{xx}-I_{zz}) = 2gN_y$$

$$I_{zz}\frac{d\omega_z}{dt} + \omega_y\omega_x(I_{yy}-I_{xx}) = 2gN_z$$

In 1771 and in 1775 Euler presented two important papers on the equilibrium and the motion of a flexible or elastic line [Euler 1771 and 1775a]. Here Euler realised that, while for flexible lines the law of equilibrium of forces was sufficient to determine the equilibrium conditions, for elastic lines, instead, the law of moment of forces was also needed. According to Truesdell,

> *in respect to the laws of mechanics this paper makes a turning point, for it is the first work on deformable continua in which the principle of linear momentum and moment of momentum appear on a par, independent and separately necessary [...] Here it is made plain that neither principle, by itself, suffices except in special cases* [Truesdell 1960:141].

Truesdell has also inferred that "Euler was led to the general principle of moment of momentum, as independent of the principle of linear momentum [...] through studies of elastic lines" [Truesdell 1968:395]. While this last statement is questionable, if one looks at Euler's view of the relationships between the principles of linear and angular momentum as it comes out from his *Scientia Navalis*, those two papers were important, since for the first time Euler could clearly separate the principles of motion he employed from the constitutive equation of the material of the line, describing its elastic properties. Again, as it had been for the rigid body in 1758, another major advance in mechanics could take place by making a clear distinction between principles of motion and the *connoissance mécanique* of bodies.

In 1765 Euler's treatise on the motion of rigid bodies, entitled *Theoria motus corporum solidorum seu rigidorum*, was published. It did not add new concepts to the matter, representing rather a *summa* of Euler's knowledge. However, the *Theoria Motus* was not to be the last work by Euler on the subject. In 1775, at the age of 68, Euler came back to the problem of the rigid body. As he had done in 1758, he devoted two papers to the matter, the first of which deals with the representation of motion, and has the role of introducing the second one. As he had done seventeen years before, Euler's goal is to separate the different aspects concerning the motion of a rigid body:

> *when it is needed to determine the motion of a given rigid body, the whole matter can be safely divided into two parts, one geometrical, the other mechanical. In the first, the goal is to represent analytically only the translation of the body from a place to another, without considering at all the principles of motion [...] this investigation belongs solely to Geometry or, rather, to Stereometry. One can readily see, however, that if this investigation is separated from the other one, belonging to Mechanics, the determination of the motion from the principles of mechanics will be much easier than if the two investigations were tackled simultaneously. Since I did so in my book on rigid bodies, the whole treatment became rather complex and intricated; I [therefore] decided here to explain more accurately only the geometrical aspects, in order to be able later to tackle the mechanical ones more easily* [Euler 1775b: 85].

Euler introduces a space-fixed frame of reference having its origin in the center of inertia of a rigid body whose axes coincide with the principal axes of the body. Let X, Y, Z be the coordinates of a point at time $t=0$. Euler's goal is to find the relationship between X, Y, Z and the coordinates of the point at time t, $x(t)$, $y(t)$, $z(t)$. He obtains:

$$x(t) = f + FX + F'Y + F''Z$$
(8) $$y(t) = g + GX + G'Y + G''Z$$
$$z(t) = h + HX + H'Y + H''Z$$

where *f, g, h* are the coordinates of the origin of frame *X, Y, Z* in the frame whose coordinates are *x(t), y(t), z(t)*. Equations (8) are the coordinate transformation between these two frames of reference. Six relationships between the nine quantities *F ... H''* can be written at once: they can be therefore represented by only three quantities (the so-called "Euler's angles"). In the same paper Euler demonstrated (only geometrically) a property that he considered *maxime abscondita*: a rigid body having a fixed point can only accomplish rotations. Today we know it as "Euler's theorem," and we state it by saying that the (orthogonal) matrix of the transformation (8) must have the eigenvalue +1.

Having solved the problem of representing the motion of a rigid body, Euler is now ready to tackle the full mechanical problem. His solution comes in the second of his papers on the subject presented in 1775, titled *Nova methodus motum corporum rigidorum determinandi* [Euler 1775c]. I will not enter into the details of Euler's method. Rather, it should be emphasized that Euler's technique of separating geometrical from mechanical aspects allows him to state, clearly and *for the first time*, the principles of linear and angular momentum:

(9a)
$$\int dm \frac{ddx}{dt^2} = gP$$

(9b)
$$\int dm \frac{ddy}{dt^2} = gQ$$

(9c)
$$\int dm \frac{ddz}{dt^2} = gR$$

(9d)
$$\int z dm \frac{ddy}{dt^2} - \int y dm \frac{ddz}{dt^2} = gS$$

(9e)
$$\int x dm \frac{ddz}{dt^2} - \int z dm \frac{ddx}{dt^2} = gT$$

(9f)
$$\int x dm \frac{ddy}{dt^2} - \int y dm \frac{ddx}{dt^2} = gU$$

where g is half of the space that a freely falling body covers in one second. P, Q, R are the component of the resultant of the external forces and S, T, U are the components of the moment of external forces, respectively. Deriving twice Equations (8) with respect to time and putting the result into Equations (9), Euler obtains six differential equations where the unknowns are f, g, h and coefficients $F \ldots H''$. Of these twelve quantities only six are independent (f, g, h and three of the coefficients $F \ldots H''$); the problem is therefore solved in general.

Truesdell has commented this momentous step with the following words:

> On every part of every body, whether punctual or space-filling, whether rigid or deformable, the total force acting upon the body equals the rate of change of the total momentum, and the total torque acting upon the body equals the rate of change of the total moment of momentum [...] These are Euler's <u>laws of motion</u> [...] They imply not only 'Newton's laws' for mass-points, but also all the other principles of classical mechanics and are just as convenient for continuous bodies as for discrete systems. [...] With Euler's memoir of 1775, the whole program of rational mechanics becomes clear. There are the two general laws, <u>common to all bodies</u> but insufficient to specify their motions. The <u>differences</u> between bodies are represented by <u>constitutive equations</u>, which specify the nature of their response to their surroundings. The constitutive equations studied in the eighteenth century are those defining the discrete system, the rigid body, the perfectly flexible line, the perfectly flexible sheet, the elastica, and the perfect fluid, with a few others that are less important. All these fit easily into the general scheme laid down by Euler in 1775. In fact, these scheme remained general enough for all of mechanics for at least 100 years [Truesdell 1968:172-173 and 128-129. See also 259-262; Truesdell's emphasis].

We may agree with Truesdell's words. However, still another point needs to be clarified. Euler, as we have seen, had been aware of the existence of two independent principles of motion since 1737, when he had conceived his *Scientia Navalis*. He had also used both principles many times, in many different papers and for several mechanical systems. However, his paper of 1755 marks the first appearance of the two principles stated explicitly together, according to their six components, written in Cartesian rectangular coordinates. Why did it take almost forty years for Euler to do so? The answer lies in his ability to separate for the first time the principles of motion from the geometry of motion. Even in his trailblazing paper of 1750, Euler had not written the two principles explicitly. The principle of the angular momentum usually appeared in his treatment after the substitution of the kinematics of the motion (i.e., the expression for the second differentials of the coordinates) into the expression of the principles of motion. What has radically changed in the 1775 paper is the way of describing the kinematics of motion, via the linear transformation (8). The importance of this change can be inferred from its effect: it is the first time that Euler is able to separate the aspects peculiar to the problem from the principles of motion so clearly that he is able to write the latter at once, explicitly, and in a generally valid way.

Epilogue

We have seen so far that the so-called cardinal laws of mechanics:

$$(10) \qquad \vec{F}^e = \frac{d\vec{P}}{dt} \quad e \quad \vec{N}^e = \frac{d\vec{L}}{dt}$$

were stated explicitly by Leonhard Euler more than 100 years after the publication of Newton's *Principia*. Why did it take so many years to state them? In this paper I have tried to summarize the most important causes, namely, the difficulty of the transition from the mechanics of mass-points to the mechanics of extended bodies, the ambiguities in the definitions of force and of the second law of motion, the inadequacy of the representation of motion and the need to separate principles of motion, valid in general, from other aspects peculiar to the problem under examination.

It must be added that "Newtonian" mechanics was not complete even after Euler's papers of 1775. As a matter of fact, a number of advances still remained to be accomplished. Only in 1780 did Euler become aware of the vectorial character of the moment of a force [Euler 1789a and 1789b]. Even more important, the mechanics of continua and the concept of stress still had to be worked out: that was to be Cauchy's great contribution to mechanics in the 1820s. We can therefore see that not only was classical mechanics not complete in Newton's works, but that it was not complete even a century after Newton's *Principia*. Thus the modern laws of motion were not "almost exactly the same" as those used by Newton in the *Principia*.

Nothing less than a gulf exists between how rational mechanics was created and how it is considered and taught today. In this paper I outlined just one case study, albeit an important one. Another case in point is the genesis of the law of angular momentum and its relationship with the law of linear momentum [See Maltese 2001: chapter 12].

Going back to the metaphors we discussed before, it seems sensible to state that teaching mechanics according to the "Apollo's arrow" paradigm is not enough. There has to be room also for the painful development of major ideas represented by the "Ancients' Inferno" paradigm. Both of these ways to present the development of science should find their place in the common practice of teaching mechanics and, more generally, science, in order to take advantage of the lesson to be drawn both from the development of algorithms and the history of ideas.

References

BENVENUTO, EDOARDO. 1991. *An Introduction to the History of Structural Mechanics, vol. 1: Statics and Resistance of Solids.* New York: Springer-Verlag.

BERNOULLI, DANIEL. 1733. Theoremata de oscillationibus corporum filo flexili connexorum et catenae verticaliter suspensae. *Commentarii Academiae Scientiarum Imperialis Petropolitanae*, vol. 6, 1732/3 (1738): 108-123.

———. 1735. Demonstrationes theorematum suorum de oscillationibus corporum filo flexili connexorum et catenae verticaliter suspensae. *Commentarii Academiae Scientiarum Imperialis Petropolitanae*, vol. 7, 1734/5 (1740): 162-174.

BERNOULLI, JACOB. 1703. Demonstration générale du centre de Balancement ou d'Oscillation tirée de la nature du Levier. *Mémoires de l'Académie Royale des Sciences de Paris* (4th ed.), 1703 (1705): 78-84. Also in: *Opera Mathematica Varia*, 930-936.

BERNOULLI, JOHANN I. 1728. Meditationes de chordis vibrantibus. *Commentarii Academiae Scientiarum Imperialis Petropolitanae*, vol. 3, 1728 (1732): 13-28. Presented in 1728. Also in: *Opera Omnia*, vol. 3: 198-210.

———. 1742. *Hydraulica nunc primum detecta ac demonstrata directe ex fundamentis pure mechanicis, Anno 1732. Opera Omnia*, vol. 4: 387-493.

CANNON, JOHN T. and SIGALIA DOSTROVSKY. 1981. *The evolution of Dynamics: Vibration Theory from 1687 to 1742.* New York: Springer-Verlag.

D'ALEMBERT, JEAN LE ROND. 1749. *Recherches sur la Précession des Equinoxes et sur la Nutation de l'Axe de la Terre dans le Systême Newtonien.* Paris.

EISENBUD, L. 1958. On the classical laws of motion. *American Journal of Physics*, vol. 26: 144-159.

EULER, LEONHARD. 1730. De communicatione motus in collisione corporum. *Commentarii Academiae Scientiarum Imperialis Petropolitanae*, vol. 5, 1730/1 (1738): 159-168. Also in: *Opera Omnia*, serie II, vol. 8: 1-6.

———. 1736. Mechanica, sive motus scientia analytice exposita. *Opera Omnia*, series II, vols. 1 and 2.

———. 1741. Dissertation sur la meilleure construction du cabestan. *Piéce qui a remporté le II. prix de l'Académie Royale des Sciences en MDCCXLI*, Paris, 1745: 29-87. Also in: *Opera Omnia*, series II, vol. 20: 36-82.

———. 1746. De motu corporum in superficiebus mobilibus. *Opusculi varii argumenti,* vol. 1, 1746: 1-136. Also in: *Opera Omnia,* series II, vol. 6: 75-174.

————. 1747. Recherches sur le mouvement des corps celestes en général. *Mémoires de l'Académie des Sciences de Berlin,* vol. 3, 1747 (1749): 93-143. Presented on June 8, 1747. Also in: *Opera Omnia,* series II, vol. 25: 1-44.

————. 1749. *Scientia Navalis, seu tractatus de construendis ac dirigendis navibus.* S. Pietroburgo. *Opera Omnia,* series II, vols. 18 e 19.

————. 1750. Découverte d'un nouveau principe de Mécanique. *Mémoires de l'Académie des Sciences de Berlin,* vol. 6, 1750 (1752): 185-217. Presented on September 3, 1750. Also in: *Opera Omnia,* serie II, vol. 5: 81-108.

————. 1751. Du mouvement d'un corps solide quelconque lorsqu'il tourne autour d'un axe mobile. *Mémoires de l'Académie des Sciences de Berlin,* vol. 16 1760 (1767): 176-227. Presented on October 7, 1751. Also in: *Opera Omnia,* series II, vol. 8: 313-356.

————. 1758a. Recherches sur la connoissance mécanique des corps. *Mémoires de l'Académie des Sciences de Berlin,* vol. 14, 1758 (1765): 131-153. Presented on July 6, 1758. Also in: *Opera Omnia,* series II, vol. 8: 178-199.

————. 1758b. Du mouvement de rotation des corps solides autour d'un axe variable. *Mémoires de l'Académie des Sciences de Berlin,* vol. 14, 1758 (1765): 154-193. Presented on November 9, 1758. Also in: *Opera Omnia,* series II, vol. 8: 200-235.

————. 1765. Theoria motus corporum solidorum seu rigidorum ex primis nostrae cognitionis principiis stabilita et omnes motus, qui in huiusmodi corpora cadere possunt, accommodata. *Opera Omnia,* series II, vols. 3 e 4.

————. 1771. Genuina principia doctrinae de statu aequilibrii et motu corporum tam perfecte flexibilium quam elasticorum. *Novi Commentarii Academiae Scientiarum Imperialis Petropolitanae,* vol. 15, 1770 (1771): 381-413. Presented on January 14, 1771. Also in: *Opera Omnia,* series II, vol. 11, part I: 37-61.

————. 1775a. De gemina methodo tam aequilibrium quam motum corporum flexibilium determinandi et utriusque egregio consensu. *Novi Commentarii Academiae Scientiarum Imperialis Petropolitanae,* vol. 20, 1775, (1776): 286-303. Also in: *Opera Omnia,* series II, vol. 11: 180-193.

————. 1775b. Formulae generales pro translatione quacunque corporum rigidorum. *Novi Commentarii Academiae Scientiarum Imperialis Petropolitanae,* vol. 20, 1775, (1776): 189-207. Also in: *Opera Omnia,* series II, vol. 9: 84-98.

————. 1775c. Nova methodus motum corporum rigidorum determinandi. *Novi Commentarii Academiae Scientiarum Imperialis Petropolitanae,* vol. 20, 1775, (1776): 208-238. Presented on October 16, 1775. Also in: *Opera Omnia,* series II, vol. 9: 99-125.

————. 1789a. De momentis virium respectu axis cuiuscunque inveniendis; ubi plura insignia symptomata circa binas rectas, non in eodem plano sitas, explicantur. *Nova Acta Academiae Scientiarum Imperialis Petropolitanae,* vol. 7:

191-204, 1789 (1793). Also in: *Opera Omnia*, series II, vol. 9: 387-398.

————. 1789b. Methodus facilis omnium virium momenta respectu axis cuiuscunque determinandi. *Nova Acta Academiae Scientiarum Imperialis Petropolitanae*, vol. 7: 205-214, 1789, (1793). Also in: *Opera Omnia*, series II, vol. 9: 399-406.

LAGRANGE, JOSEPH LOUIS DE. 1888. *Mécanique Analytique*. Paris : Quatriéme edition.

MACH, ERNST. 1968. *La Meccanica nel suo sviluppo storico-critico*. Turin: Boringhieri.

MALTESE, GIULIO. 1992a. *La storia di "F=ma". La seconda legge del moto nel XVIII secolo*. Biblioteca di Nuncius n. 7, Florence: Olschki.

————. 1992b. Taylor and John Bernoulli on the vibrating string: aspects of the dynamics of the continuous systems at the beginning of the eighteenth century. *Physis*, vol. XXIX: 703-744.

————. 2000. On the Relativity of Motion in Leonhard Euler's Science. *Archive for the History of Exact Sciences*, vol. 54: 319-348.

————. 2001. *Da "F=ma" alle leggi cardinali del moto: sviluppo della tradizione newtoniana nella meccanica del '700*. Milan: Hoepli.

NEWTON, ISAAC. 1687. *Principi matematici di filosofia naturale*. Italian translation by A. Pala, Turin: UTET, 1977.

RAVETZ, J. 1961. The representation of physical quantities in eighteenth Century Mathematical Physics. *ISIS*, vol. LII: 7-20.

SCOTT, WILSON L. 1959. The Significance of 'Hard Bodies' in the History of Scientific Thought. *ISIS*, vol. L: 199-210.

SPEISER, DAVID. 1983. The principle of relativity in Euler's work. In *Symmetries in Physics (1600-1980)*, M.G. Doncel, A. Hermann, L. Michel and A. Pais, eds. 1st Int. Meeting on the History of Scientific Ideas, Saint Felin de Guixals, Catalogna, Spain, September 20-26, 1983.

TAYLOR, BROOK. 1713. De motu Nervi Tensi. *Philosophical Transactions of the Royal Society of London*, vol. 28: 26-32.

TRUESDELL, CLIFFORD AMBROSE. 1955. Rational fluid mechanics 1687-1765. In *Leonhardi Euleri Opera Omnia*, series II, vol. 12, part I: I-CXXV.

————. 1960. The Rational Mechanics of Flexible or elastic bodies, 1638-1788. In *Leonhardi Euleri Opera Omnia*, series II, vol. XI, part 2.

————. 1968. *Essays in the History of Mechanics*. New York: Springer-Verlag.

WILSON, CURTIS. 1987. D'Alembert versus Euler on the Precession of the Equinoxes and the Mechanics of Rigid Bodies. *Archive for the History of Exact Sciences*, vol. 37: 232-273.

A Historical Survey of Impact Theories

Piero Villaggio[1]

Impact between solid bodies constitutes one of the most fascinating branches of mechanics. Several models have been proposed for describing it in the last 2,400 years, but none is completely satisfactory, even among those formulated in the last decades. Though many debated questions have been clarified, there are still aspects of the collision which require a rigorous treatment.

The emergence of the concept of impact

Impact has been observed by men since their appearance on Earth. While many natural phenomena, such as the motions of heavenly bodies, the alternation of day and night and the changes of seasons, appear perfectly regular and exactly predictable, some other phenomena, such as thunder, avalanches and wind squalls, are occasional, violent, abrupt and, above all, accompanied by strong, often destructive, forces. But the possibility of generating large forces by collision has been also exploited by men since the Stone Age to break stones, split trunks, drive piles in the soil, hunt animals for procuring food and other numerous domestic operations. And even many millennia before men, monkeys knew how to smash the shells of some fruits by simply throwing them from the branches of tall trees (Figure 1).

Figure 1. Monkeys knew how to smash the shells of some fruits simply by throwing them from the branches of tall trees

[1] Università di Pisa, Dipartimento di Ingegneria Strutturale, Via Diotisalvi 2, 56126 Pisa, ITALY

But centuries of subsequent observations and research have proven that impact phenomena are much more diffuse and articulated than they appear at first sight. The impact of an ocean wave against the keel of a ship has the same effect as the impact of a stone; the explosion of a mixture of easily-inflammable gases generates in the air a shock wave having exactly the same mechanical properties as a water wave; the rapid propagation of an electrical charge may produce the same destructive consequences as the fall of a serac. At the same time, the technical applications of impact can be found in an unbelievable variety of uses: hammering, forging, cutting, beating skins and cereals, these have been become the fundamental operations of all kinds of labour since the beginning of the Neolithic Age.

Impact is a branch of mechanics, like statics, particle dynamics and systems dynamics, but, if we consider its collocation in all treatises of mechanics, we note with surprise that the subject is regarded as atypical. A separate chapter is reserved to it, new axioms are introduced, another – more elementary – mathematics is used in order to solve problems involving colliding bodies. And the fact is even more disconcerting when we compare these simplistic approaches with the other sophisticated and elegant treatments of problems such as that of the three bodies or of the motion of the gyroscope, which, in principle, should be easier. And the same disturbing dichotomy persists in the applications. Today it is possible to evaluate almost exactly the stresses in a huge structure such as a ship or an airplane, but many everyday problems in which collision occurs still lack a precise formulation. Consider, for instance, the problem of driving a thin iron piton into a soft substance (Figure 2).

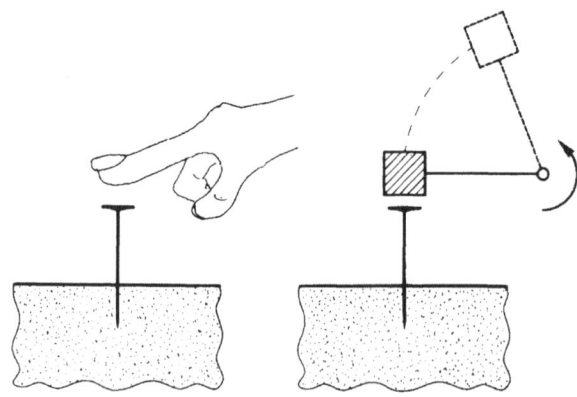

Figure 2. Driving a piton into a soft substance

Sometimes we perform the operation by simply pushing the piton with a thumb; sometimes we use a hammer. If we try to formulate each of these operation in terms of equations we immediately see that both are difficult, but the second one is by far harder. But the more surprising consideration is that every workman knows which is the best way of placing a nail, either statically or impulsively, simply by evaluating the toughness of the medium and the slenderness of the nail.

Returning to the rigorous theory of impact, let us read the introductory description of the collision between solid bodies as presented by W.J. Stronge:

> When a bat strikes a ball or a hammer hits a nail, the surfaces of two bodies come together with some relative velocity at an initial instant termed incidence. After incidence there would be interference or interpenetration of the bodies were it not for the interface pressure that arises in a small area of contact between the two bodies..., the pressure in the contact area results in local deformation and consequent indentation... [Stronge 2000: 1].

The explanation sounds perfectly precise, but after a bit of meditation, some subtle doubts arise. The indentation between two colliding bodies in not an exclusive characteristic of collision, for, as in the case of slow mutual compression between two convex surfaces, the initial, infinitesimal, region of contact starts to gradually expand until the local stresses are able to balance the final force of compression. At the same time it is said that, after this phase of mutual approach, a period of restitution begins, in which the elastic part of the internal energy is released, driving the bodies apart. But this is also what happens when we push two tennis balls together and then remove the compression action, slowly distancing their centres. The borderline between impact and gradual contact is not well defined. We can establish with certainty the extreme situations, and say, for example, that the collision of a cannon ball against a wall of a fortress is a case of impact, while the squeezing of a rubber balloon with a finger is a case of smooth contact, but how to classify many other instances, like the fall and the splashing of a drop of mercury in a viscous oil? A possible criterion of distinction may be that of calling 'impact' all the cases of contact between rigid or deformable bodies when the inertia terms prevail over the other forces, but the rule is not very sound, for it would regard as impact the uniform flow of an airplane in the atmosphere or that of a ship on a calm sea.

The attempts at defining impact

There is deeply-rooted conviction that, since the beginning of natural science, philosophers tried to explain the action of impact according to mechanical principles. The best-known argument is a passage from the *Questiones*

Mechanicae, attributed to Aristotle, in which the author asks himself why it is possible to split a large, heavy body with a small and light wedge. The answer is, because here again, as in statics, the principle of the lever applies, provided that forces are replaced by the products of weights by their respective velocities. The explanation is "not very convincing" [Dijksterhius 1986: 32], as it contains a new extension of the concept of force, but it shows an effort at treating impact in terms of ordinary mechanics, which will persist for eighteen centuries after Aristotle. This aptitude can be named "Horror collisionis".

The onset of sudden changes in the regular motion of a body was congenially extraneous to Greek thinkers. They were concerned with invariable, eternal, laws of nature; every irregularity was ignored or regarded as a small, occasional, disturbing, perturbation. Heavens rotate with a perfect geometric order; a constant force causes a constant velocity; natural motions must be circular; collisions are easily explainable with the law of levers; technical applications do not belong to science!

This aristocratic mentality, protracted for centuries, did not prevent engineers, architects, headmasters, generals and admirals from exploiting the formidable potential power of impact for their technical (and military) purposes. It is only at the end of the sixteenth century that official science began to consider impact as an important branch of mechanics as it offered the conceptual instrument for describing the terminal ballistic of a cannon ball as well as the effect of a forging hammer on a piece being worked. At that time these topics were at the forefront of dynamics for warfare, technology and navigation. The debate on the nature and measure of impact represents a vivid picture of the difficulties that had to be overcome before the phenomenon could be properly disentangled. The precise questions on which scientists of the time were obliged to decide were essentially two [Maltese 1992]: What is conserved during impact, force or momentum? Is impact an instantaneous phenomenon or a continuous process of deformation and restitution? According to a scheme introduced by David Speiser in a course of lectures delivered at the Scuola Normale in Pisa in 1997, the four questions can be inserted into a rectangle and each term of the upper part can be combined whichever term of the lower part, offering the choice between four theories, as for example, conservation of force and continuity or conservation of momentum and continuity, and so on (Table 1).

Force	*Momentum*
Inst.	*Contin.*

Table 1

It is very interesting to notice that the scientists who considered the problem, such as Galileo, Descartes, Huygens, Leibniz, Newton, Euler and many others, have given different answers with ingenious and convincing arguments.

The unification of the theories

As has often happened in the history of science, the contradictions between the four theories were more apparent than substantial, and their reconciliation achieved, not by inventing a new first principle of mechanics, but simply by exploiting the (at that time) hidden properties of differential calculus. For scientists of the seventeenth century the solution of a differential equation had to be necessarily a smooth function, graphically representable by a regular curve. Sometimes they ideally decomposed a regular trajectory into a sequence of small impacts, as Newton does for explaining the effect of the centrifugal force [Maltese 1992: 19] or as Johann Bernoulli does in his analysis of the sliding of a block on a moving triangle. But these operations are only technical artifices in order to reconstruct a trajectory that should be necessarily regular.

The date of birth of the reconciliation of the contrasting theories of impact is the year 1712, when B. Taylor proposes a partial differential equation for describing the transversal vibration of a string. The equation is:

$$\frac{\partial^2 u}{\partial t^2} = c^2 \frac{\partial^2 u}{\partial x^2} = 0$$

where c is a constant. There is no agreement between the historians of mechanics whether Taylor did effectively write this equation. Szabó claims that he 'unconsciously' did [Szabó 1979: 320]; Truesdell says that after a "brilliant beginning he gets lost in a morass of special assumptions and errors" [Truesdell 1960: 130]; mathematicians of the following generation would object that a simple differential equation, without boundary and initial data, is insufficient for a correct formulation of the problem. But let us set these reservations aside and accept the year 1713 as the conventional start of partial differential equations. In 1747 J.B. d'Alembert finds a general solution of the equation of a vibrating string of the form

$$u(x,t) = p(x - ct) + q(x + ct),$$

where p and q are arbitrary functions. D'Alembert, however, does not realize that p and q can be functions admitting corners, and considers analytic forms for p and q. It is Euler in 1748 who notices that p and q can have piecewise continuous first derivatives [Szabó 1979: 335].

Euler's remark is a milestone in impact theory. A softly deformed string and a violently locally-plucked string are ruled by the same differential equation, provided that we extend the class of solutions. This implies that there is no need of a separate treatment between smooth and impulsive mechanical theories once we introduce some mathematical generalizations.

The full exploitation of the potential properties of d'Alembert's solution had to wait for more than a century until Saint-Venant (1867) applied d'Alembert's method to an elastic rod, struck longitudinally at its top by a falling mass (Figure 3), which, for a certain period of time, remains attached to the top before rebounding.

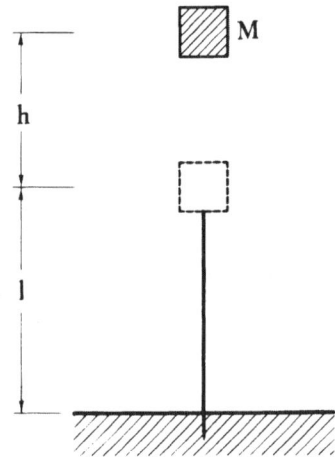

Figure 3. D'Alembert's method applied to an elastic rod struck longitudinally at its to by a falling mass

Saint-Venant's solution represents one of the most skilful exploitations of mathematical physics, since the author is able to construct the solution taking into account the subsequent reflections of the elastic wave front emanating from the upper end of the bar at the instant of the first contact. Saint-Venant's solution is purely one-dimensional and hence leaves open the question of the influence of the transversal dimensions of the cross-section of the bar on the process of wave propagation. This interrogative was answered a few years later by F. Neumann (1885), who found an elegant solution for the longitudinal mutual impact of two elastic bars having circular cross-sections of equal diameters. In contrast to Saint-Venant, Neumann's solutions includes the radial displacements in addition to the axial ones, thus explaining that, even in case of perfectly elastic materials, a fraction of the initial kinetic energy possessed by the colliding bars is instantaneously converted in exciting the radial motion in them so that the rebound is never complete. This phenomenon of apparent dissipation is usually accompanied by another apparent dissipation due to the local indentation occurring in correspondence of the regions of first contact. This localized additional state of displacement is difficult to determine, but in the special case of coaxial impact of an elastic sphere against the domed end of a slender rod (Figure 4), the local indentation can be exactly evaluated by using a celebrated elastic solution due to H. Hertz (1882).

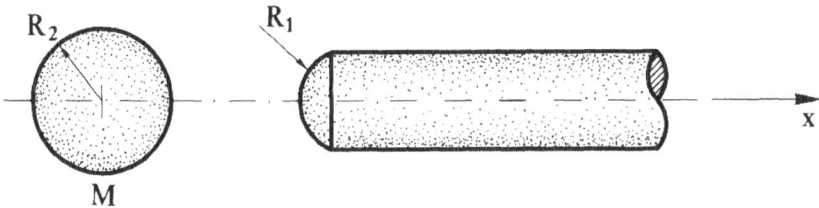

Figure 4. Coaxial impact of an elastic sphere against the domed end of a slender rod.

The efforts spent in solving d'Alembert's equation are concentrated on the particular problem of longitudinal impact, but what can be said in the case, perhaps even more important, of the transverse impact of a beam? (Figure 5.)

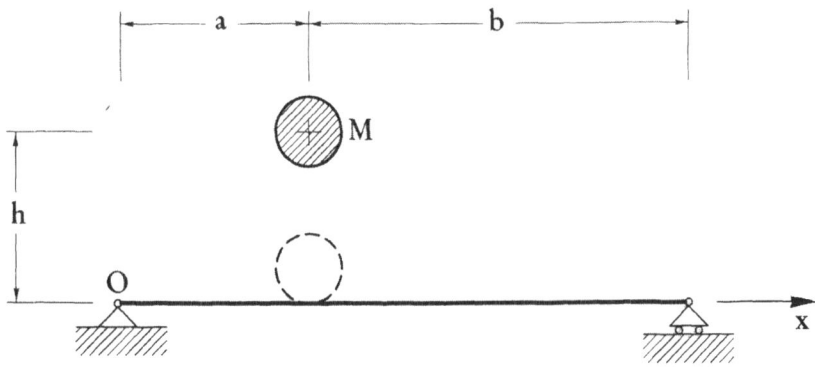

Figure 5. The transverse impact of a beam

The answer is disappointing. The dynamic equation of the beam is of the fourth order in the x-variable and of the second order in the time variable. It does not admit the beautiful solution $p(x+ct)+q(x-ct)$, which includes singular initial data. The only possible device is to expand the solution and the initial data into series of known functions whose coefficients must be determined. But, besides the difficulty of interpreting solutions under form of series, the delicate question of the convergence of these series arises, because singular data usually yield divergent series. The argument is still object of study [Weinberger 1965].

The approximate methods of solution

Exact solutions are complicated and limitative, unable to tolerate the least extensions. For example, if the density of a string or a rod is variable, explicit solutions are hardly obtainable. On the other hand, in physics and engineering there are innumerable cases of impact between three-dimensional bodies whose surfaces are endowed with sharp edges or tips.

In the face of these situations scientists have assumed a different attitude. Is the detailed information contained in the exact solutions really necessary? When a locomotive enters a bridge at high velocity, elastic transverse waves arise that go to and fro between the ends of the bridge. But often these facts are superfluous for an engineer who wishes only to know the localization and the magnitude of the highest stresses in order to guarantee the safety of the structure. And he only tries to evaluate these quantities through an easier, though approximate, procedure. This unprejudiced position has given birth to the so-called 'semi-statical' methods of dynamics, based on a crude equivalence between the kinetic energy of the impinging mass and the strain energy, statically determined, at the end of the compression period when the struck body (bar or beam) is instantaneously at rest. Knowledge of the strain energy is often sufficient for determining the highest stresses, provided that we have a reasonable idea of the points where they arise. This ingenious method has been introduced, before the appearance and diffusion of Saint-Venant's results, by merit of Cox (1849), an engineer appointed to inquire the application of iron to railway structures.

But, in this case again as often in the progress of mathematical physics, even a dirty procedure like that introduced by Cox has been soon justified in the ambit of a general theory. The leading idea is simple. In absence of dissipation, the total energy, sum of potential and kinetic energies, is conserved. Hence its first variation is zero, or, in other terms, the total energy is stationary. The exact evaluation of this point of stationariness yields the solution of a difficult partial differential equation. Yet we may content ourselves with approaching this critical point by linear combinations of easily manageable functions, hoping that the result is a good approximation of the exact solution. This procedure is now universally known as the Rayleigh-Ritz method. Of course the question arises of the convergence of the method and this is one of the most delicate problems studied by mathematicians of the twentieth century.

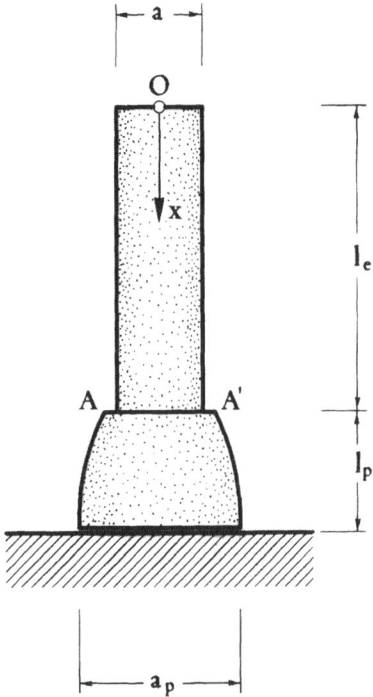

Figure 6. The plastic squashing of the front of a projectile fired against armour

The influence of nonlinearities

There is a second milestone in the theory of impact. Though difficult, d'Alembert's equation and its generalizations are linear and hence enjoy of the privileged properties of linear equations [Weinberger 1965: 29]. But suppose that we want to describe mathematically the longitudinal motion and change of pressure in a tube containing a gas. After some manipulation we arrive at an equation almost identical d'Alembert's equation, but with the difference that now the unknown u represents a density instead of a displacement and that the coefficient c, which is not a constant, is a function of this density. Since the density is just our unknown, the equation is nonlinear. Now B. Riemann (1860) finds the astonishing result that, even starting with a very smooth initial distribution of density, this density can become discontinuous after a certain time. Riemann's discovery signs the birth of the theory of shock waves is gas dynamics. Supersonic aerodynamics, the propagation of explosion and the onset of tempests in the atmosphere, are still ruled by Riemann's equation.

As it has happened for d'Alembert's equation, many efforts have been devoted to extending Riemann's solution to the three-dimensional case. The question is not purely theoretical, but also practical as, in general, the wave fronts are surfaces moving in space. In this case again an explicit solution is not known, but some significant information can be derived by reversing the problem. Assume that in a gaseous medium there is a travelling shock front. In general, density, pressure and temperature are discontinuous across the front, but certain combinations of these quantities are continuous. The analysis of these conditions of continuity, initiated by Rankine (1870) and Hugoniot (1887) has brought new important knowledges in the physics of gases.

The onset of shock waves is not necessarily restricted to compressible fluids, since also in solids surfaces of discontinuity can appear where the constitutive properties of the material changes as a consequence of an excessive state of stress. The most typical instance is that of the plastic squashing of the front of a projectile fired against armour (Figure 6). G. Taylor (1948) has proposed an elegant theory for determining the permanent deformation of a bar impinging vertically a rigid wall. Taylor's theory – perfectly confirmed by experiment – applies Rankine-Hugoniot's conditions at the section AA' of separation between the elastic and the plastic regions.

Shocks in the theory of combustion

The heat propagation is governed by the equation (proposed by Fourier, 1822)

$$\frac{\partial u}{\partial t} - k \frac{\partial^2 u}{\partial x^2} = 0 \, ,$$

where u is the temperature and k a constant. In contrast to d'Alembert's equation, the heat equation has the property that, even if the initial datum is piecewise continuous, temperature becomes immediately smooth after the first instant. This fact reflects the circumstance that temperature is an equilibrium concept, that is, the rate of change of it must be slow relative to the time scale of random molecular motions that produce equilibrium. But this universally-accepted belief has been questioned in the study of heat propagation in the presence of a thermal source, for example, a combustible substance. In this case, Fourier's equation must be replaced by the equation

$$\frac{\partial u}{\partial t} - k \frac{\partial^2 u}{\partial x^2} = u^p \, ,$$

where p>1 is a constant. Fujita (1966), the first to study this equation, has found the surprising result that the blowing up of the solution depends on a critical value p_{cr} of the exponent p: for $p < p_{cr}$ the solution blows up after a certain time; for $p > p_{cr}$ the solution remains bounded provided that the initial value is sufficiently small [Levine 1990]. Results of this kind are clearly not only interesting to mathematicians, but also to workers in diverse areas such as chemical reactor theory and combustion theory. For example, the knowledge of p_{cr} may be essential for determining the temperature of ignition of the fuel in an engine in order to optimise its efficiency.

But the applicative utility of these results are even more general. It has recently been observed that the velocity of diffusions of some epidemic diseases is governed by a blowup equation. And this explains why diseases, stable for decades, suddenly explode once hygienic or climatic conditions degenerate below a certain threshold.

New extensions of impact theories

Traditionally scholars, accustomed to regard impact as a particular branch of rigid body mechanics, are reluctant to accept that phenomena of impact occur in deformable solid bodies, provided that we enlarge the definition of impact as that in which 'solutions become singular in time'. They are even more restive in recognizing that similar phenomena arise in fluid mechanics and even in heat propagation. But, if we admit that a precise mathematical definition is safer than others based on nebulous descriptions, then the entire theory of impact becomes more rigorous and accessible.

In this perspective it is not surprising that biologists and economists are inclined to apply impact theories in their disciplines. A zoological species survives for millennia but sometimes extinguishes within few centuries; is not this a case of impact provided that we contract the time scale? Again, a market seems florid and stable, but suddenly stocks collapse causing a bankrupt; is it not possible to write the equation of the failure? These problems, a few decades ago seen as far-fetched, are now the object of intense studies.

References

DIJKSTERHUIS, E.J. 1986. *The Mechanization of the World Picture*. Princeton: Princeton University Press.

LEVINE, H.A. 1990. The Role of Critical Exponents in Blow up Theories. *SIAM Review* vol 32, no. 2, 262-288.

MALTESE, G. 1992. *La Storia di "F=ma". La seconda legge del moto nel XVIII secolo*. Biblioteca di Nuncius, n. 7. Florence: Olschki.

STRONGE, W.J. 2000. *Impact Mechanics*. Cambridge: Cambridge University Press.

SZABÓ, I. 1979. *Geschichte der mechanischen Prinzipien*. Basel: Birkhäuser.

TRUESDELL, C.A. 1960. *The Rational Mechanics of Flexible or Elastic Bodies 1638-1788*. Introduction to *Leonhardi Euleri Opera Omnia*, second series, vol. IX (2). Zürich: Orell Füssli.

WEINBERGER, H.F. 1965. *A First Course in Partial Differential Equations*. Waltham: Blaisdell.

The Gothic cathedral of Strasbourg

WHAT CAN THE HISTORIAN OF SCIENCE LEARN FROM THE HISTORIAN OF THE FINE ARTS?

David Speiser[1]

Certain aspects of and approaches to historiography, such as the importance of commissioned works and the relationships between related disciplines, were already used by the historians of the fine arts at a time when there were only very few historians of science. Benvenuto and Truesdell, both great lovers and connoisseurs of the arts, taught us that science, like art, is beautiful, and that its history, when presented by great men, can be beautiful as well.

Introduction

Even today the possibilities for learning the craft of the historian of science are not abundant; when I was a student, they hardly existed at all. In my case I learned what I could through a few, albeit very important, personal contacts, and one of these teachers I am invited to honour today.

But then I was also an avid visitor of art museums, churches, palaces, etc., and I read books and followed lectures on the history of the fine arts. One contact between the history of science and the arts themselves is immediate: often we can note that works of art prove that precise scientific and technological knowledge already existed at a very early time, long before a continuous and systematic development of science had begun. This knowledge was acquired by artists and artisans intuitively when they exercised their craft. Such testimonies are an essential part of the prehistory of a certain field of science, and the history of that field would not be complete, without an account of them.

One example are the pyramids built during the old Egyptian empire. They testify to the powerful technology of their builders, but also to a geometric ideal shared by the sculptors of the same period. This must be seen together with how the statues express a new appreciation of three-dimensional space and of plasticity as an aesthetic value in itself. We may even see the spherical wigs that the ladies of that time wore as a compliment to that artistic ideal. Towards the end of the second millennium, we find the famous ornaments of the new Egyptian empire analysed by Andreas Speiser [1927]from the point of view of mathematics. He pointed out that all tetragonal space groups had been found by the Egyptians, including one denoted C(4v)II, which is a highly nontrivial one, a

---

[1] Prof. Emeritus, Catholic University of Louvain, Bromhübelweg 5, CH-4144 Arlesheim, SWITZERLAND.

fact that testifies to a systematic occupation with these geometric questions. This art was later developed further by the Arabs.

Another example are the circular medallions found in the so-called Tomb of Agamemnon in Mycenae, from the end of the seventeenth century BC [D. Speiser 1983]. This well-known figure, composed of six leaves, shows that its artisan had grasped, at least intuitively, a geometric theorem, perhaps the first one discovered in Europe.

Nothing needs to be said of the role of the arts for the genesis of Greek science and of the artists of the Italian Renaissance for modern science: how many artists of those periods were not scientists and vice versa? [Speiser 2000].

But there remains the puzzling question of the state of science in the Middle Ages as testified by the great medieval buildings, especially the cathedrals and their towers, the highest towers built before the middle of the nineteenth century? That these works of art testify to an extraordinary engineering skill on the part of their builders is manifest, but is the beauty of these monuments also connected to their science? I mention this only because I shall return to this question later. But first I will address myself to the relation between historiography of the arts and historiography of science, which is my main theme.

Quite late in my activity as a historian and only slowly, very slowly, did it dawn on me that in spite of the fact that science and the fine arts are radically different endeavours of the human mind, the ways and means by which their historiographies proceed are often remarkably similar. And for reasons that are, for many of you, obvious, but will become clear to all during this paper, this symposium seems the appropriate place to consider these parallels.

I shall, of course, restrict myself entirely to the historiographies of the fine arts and the sciences; only at the end shall I say a few words about science itself.

The task of the historian and his ground work

For his work the historian of science first of all finds before him publications of the scientist, his articles and books, which present the scientific discoveries and also the deepening comprehension of earlier ones, which both constitute the development of science. Behind these documents the historian can feel and sometimes even see, more or less clearly, the scientist who wrote them. Sometimes he also can see the context or line of tradition in which the scientist stands: his teachers, the books and articles that he read. More often he knows the disciples, students or other reader of a scientist's works, who formed what is sometimes called his 'school', and more generally the historian can often discern the scientist's 'sphere of influence', which expands for a certain time and then shrinks again. Sometimes it is more appropriate to focus on and speak of a team, today more so than ever. But teamwork has a long tradition in science, especially in the applied sciences: one has only to think of the architects and of the builders of bridges, fortresses and of ships, etc. These activities proceed side by side with interaction with developments in other, related fields, arithmetic, geometry, mechanics, chemistry, architecture, engineering, etc.; some of these interact with each other quite intensely. Sometimes research, especially of an applied activity, is initiated by the scientist but more often it is commissioned.

In short, we would say that the historian of science works with:

- the works of science

- the scientists themselves

- their teachers and disciples

- the colleagues and contemporaries with whom they collaborate

- interaction with other related fields of discipline

- investigations initiated by the scientist or commissioned of him by a third party

To each but one of these points we find a counterpart in the historiography of the arts, namely,

- the works of art

- the artists themselves

- their teachers and masters as well as their disciples in the 'atelier'

- the team, especially in the architects firm

- commissioned works of art versus those that are due to the initiative of the artist.

But we must note that all these aspects and approaches were followed and developed by the historians of the fine arts, and very systematically so, at a time when there were only very few historians of science.

The role and importance of commissioned works

To begin with the last point, I think it is obvious that the great majority of works of art from the pyramids until at least 1800 have been commissioned: churches, altarpieces, palaces, villas and their decorations. Many of them must be counted as great works of art. Furthermore monuments and frescoes were mostly not created on the initiative of an artist, but were, and still are, commissioned by a patron, and every art historian is aware of this.

Much the same holds for the development of science as well, but here the importance of commissioned research is in general not stressed enough. Again, think of the pyramids and the impact which their construction must have exerted on the then still rudimentary geometry and mechanics; think of the construction of the medieval cathedrals and towers to which I shall come back; and recall Galileo's praise of the Arsenal of Venice at the beginning of his *Discorsi*. But even in modern times this impact is often underrated; I remind you of cartography, the making of precision clocks, river corrections, the chemical industries and of the great laboratories during the second World War: Los Alamos, the Radar Lab and the Sonar Lab. Clearly all of these had in many ways a decisive impact not only on applied but also on fundamental science. And I need not remind you of today's enormous efforts in the field of medical research, be they now commissioned by the state or guided by an enterprise such as a pharmaceutical company. These are all decisive pieces of the history of science and often very potent stimuli, but again, all too often they remain underrated.

The mutual stimulation between branches of science

Another point where we can learn from our colleagues in the fine arts concerns the relationships between the various domains of science. To show this I will use examples of my own science, physics. It often strikes me how little many physicists and even historians of physics are aware of how much physics owes to its sister sciences, mathematics and chemistry.

That modern physics is inconceivable without the discovery and invention of the infinitesimal calculus has often been stated, but even so not always enough emphasis is given when one discusses the details of the progress made. If a scientist is classified as a mathematician, his contribution to physics may be undervalued, even if his contributions to mechanics and to physics fill far more volumes than his contributions to mathematics. I am speaking here of Euler, whose contributions to physics is slighted or even overlooked altogether by physicists and their historians. Much the same can be said of the work in mechanics of James Bernoulli and of Cauchy. We needed Clifford Truesdell to show and to explain to us the fundamental importance for mechanics and physics of the works of all three of them. On the other hand, this is a common problem. I am often surprised to see how much mathematicians and *their* historians underrate the stimulation their science received from other branches.

Likewise, historical accounts of physics often overlook the fact that after Dalton the real champions of atomism were the chemists: think of Kékulé and those who created stereometric organic chemistry, at a time when only a minority of physicists believed in atoms; some physicists even opposed atomism violently as late as the beginning of the last century. I need not say how little of all this arrived into the mainstream of the historiography of physics.

Now compare this state of affairs to a historiography in the fine arts, say a book about sculpture in the Renaissance. If this book did not present and discuss carefully the influence of related fields such as painting and architecture, it would be rated as a very poor book indeed.

The scientist in his tradition

What is the state of affairs of our efforts to determine the intellectual roots of a scientist, the sources of his knowledge and, even more important, the efforts to find out what mainly stimulated him? And what is the state of our knowledge about a scientist's influence on those who came after him? In these cases I feel that we may claim justly that the historians of the arts have an easier time of it than we do: why?

This, I believe, is due mainly to the overwhelming number of original scientific publications: books, papers, letters, etc., which makes it extremely difficult to keep track of the transmission of knowledge through the generations and which makes *bona fide* errors almost unavoidable. Here is an example: I was told, or perhaps I myself concluded from what I was told, that Newton had learned Kepler's laws from the *Astronomia Nova* and from the *Harmonice Mundi*. Then almost accidentally I learned from Bernard Cohen, the well-known Newton scholar, that Newton had never read these books, but had learned the laws from a minor publication. Likewise, Newton in all probability never saw the *Discorsi.*

Thus, to show that someone had read a certain work of a predecessor is never as easy as it looks, and it is often impossible to prove that, even if he had read a work, he received an idea from it. Surely the way the art historians can document these links is impressive, but I would not be so bold as to tell a young historian of science that to find these links is a foremost priority for serious research, and I would warn a young scientist to be careful with his time, interesting and even haunting or tantalizing as the question may be.

The historian's principle object and his aim

But let me now come to the most important, indeed, the central comparison in this list of parallels between the historiographies of the fine arts and the sciences: the comparison between their respective relations to their immediate object of research, the work of art on the one side and the scientific publication or as the case may be, the experimental setup on the other.

Historiography of the arts is inconceivable without an intimate, not only a visual, relationship to the work of art, and I suspect that in most cases such a relationship must mature over a long period of time. Every art historian knows that the works of art and their contemplation are the *raison d'être* of his whole profession, indeed for most of them it is the principal motivation for their intense and often passionate relationship with the history of the arts. With respect to the historiography of science we can observe over and over again that this is not necessarily the case at all.

Of course, the art historian has it much easier here ... at first: works of art are directly accessible to him, just as they are to everybody, especially in museums but even on our public streets and squares. Therefore it is also easier to penetrate into them and to gather facts concerning them. It is precisely this 'easy access' that the historian of science lacks. The older scientific documents are not easy to find, let alone to understand, as everyone who has tried will agree. Thus for understanding and inserting into their proper historic context the old as well as the new documents, one must have been educated and trained in the specific scientific domain being studied, whereas the art historian need not be a professional artist.

For this reason alone I am convinced that the most important task in the field of the history of science is to edit the old documents and to make them accessible for scientists through introductions and comments. I want to underline here that both of the scholars whom we honour today did this in an exemplary way.

Here an important lesson awaits all scientists who care for the history of their own field, and thus I shall appeal to you. It is inconceivable that anyone should write about an artist without having thoroughly looked at his works and having pondered them. Less than a month ago the art critic of the *Herald Tribune*, Souren Mellikian, whom some of you may know, wrote on the occasion of an exhibit entitled "Women in Rembrandt's Life", that while, except for the three or four women most closely connected to the painter, little was known about the women whom he painted, a careful look at the faces in the portraits and etchings would reveal Rembrandt's great psychological sensitivity for each individual and yield to the spectator deep insights of each one. And indeed the same holds true for scientists.

The physicist Arthur Wightman once told me that André Weil, who, when he wrote his book *The History of Number Theory*, had studied extensively the works of Euler, said in a private lecture, "after having now penetrated into Euler's work on number theory, I think that I know him better than I know most of my best friends!"

Even the writer of a *biographie romancée* that is only flimsily connected to its subject, takes a good, if perhaps mostly sentimental look, at the works of his artist. But all too often, I am afraid, this does not hold for the scientists.

I am sure that all of you have heard anniversary lectures delivered by a scientist, perhaps even at a university, where the speaker had manifestly not bothered to lift the cover of even one book of the famous man whom he was invited to honour. Rather he was satisfied to tell his audience what he had learned from the footnotes he had found in the textbooks read during his student years. Or perhaps he had heard it only in one of the lectures he attended (his own?). To act so with respect to scientific matters proper is inconceivable, but concerning the history of science everything seems to be permitted – and accepted!

Here I must appeal to all of you to do all in your power to prevent such unworthy performances. They are not only indecent, but they do great harm to the cause of the history of science as a rigorous academic discipline, to both its research and its teaching: in short, they are counterproductive! Small wonder that so often the history of science is considered a mere curiosity. You know that there are many universities where history of science is not a regular discipline, and if such fraudulent and swindling speeches are given, the authorities, and even more importantly, our dear colleagues, can say, "why subsidize chairs and even institutes, when each professor can do this for free?"

And that is not all, of course. The history of science claims, rightfully I think, to broaden the horizons of its students and of all interested persons, scientist or not. And again it is not the least merit of the men whom we are honouring today that they did just that. Thanks to both of them we have now a history of science richer than ever before, and which did indeed open new horizons.

Before saying a few words about the men we honour today, I must add a point, a main point indeed, thanks to which the historian of science often has it easier than his colleague in the arts. With respect to the arts the German historian Ranke's word that "each period stands in direct and immediate relation to God" is valid; there is no progress. While I am not saying that after the Italian Renaissance no equally great works of art were produced, I am certain that during later periods, no works *greater* than the most beautiful ones of the Cinquecento were ever made.

Not so in science. There we can notice and establish progress. By this I do not mean the mere accumulation of vaster and vaster amounts of detailed information, let alone a general increase of the quality of research. Rather, through time science establishes more and more connections between more and more phenomena in each field, and through the discovery of new concepts and new laws eventually it creates larger and larger, connected, domains. Especially in the exact sciences, thanks to the newly-discovered and rigorously-developed concepts and to the theorems formulated in mathematical language what we have learned by observation, or experiment, by imagination and deduction, becomes unified. These unifications of the exact sciences through general theories are, I guess, the best measure of this progress. This stands in striking contrast to what we see in the history of the arts with its upsurges and downfalls, where progress exists only in a subordinate and relative way.

As a consequence of this progress through unification the historian of science can always say to even the greatest scientist who preceded him, "while I admit with great respect and pleasure that you penetrated more deeply into nature than I do, I can now see more clearly than you could, what you were up to!" No art historian, nor even an artist, can say this. This possibility, which is based on the new perspectives offered by the progressing unification, makes our task easier. It is also a great relief for the historian; it is certain that in the not-too-distant future, historians will see many things even more clearly than we can see them today.

It follows that history of science will never give a final and absolute account; its account can always be improved upon. Thus the historian must be judged according to the words of the German poet Friedrich Schiller: "He who has fully satisfied the best of his contemporaries has lived for all times!" In this spirit we must honour the two men, Clifford Truesdell and Edoardo Benvenuto, about whom I shall now say a few words.

Benvenuto: mechanics and the art of restorations

I will begin with Edoardo Benvenuto, whom I knew much less well: once I was his guest and he showed to my wife and me the beautiful churches and *palazzi* of Genoa! But except for the conversation during the meal to which he invited us, most of what I know about his views and ideas I gathered from copies of papers sent to me by Profs. Foce and Corradi, to whom I am much indebted. I learned from at least two of these works that our ways were strangely intertwined; we dealt at least twice with the same subject and both times each of us was ignorant of the other's efforts, which now, of course, I regret.

The first subject with which we both dealt, I on my side in collaboration with Mrs. P. Radelet, Anne de Baenst-Vandenbrouck and J.L. Pietenpol, was Daniel Bernoulli's first paper on mechanics, where he deals with the 'parallelogram of forces' [D. Speiser 1987]. This paper opened a long series of axiomatic, as we say today, investigations on the law of the composition of forces, thus on the concept of force. Benvenuto carefully follows the long story of these investigations, from which much can be learned [Benvenuto 1985]. Mrs. Radelet's and my own aim, when, some time later we edited the third volume of the complete works of Bernoulli, was more modest: we just wanted to explain in detail what he did and why, and indicate some later criticisms. I regret, not so much that I could not discuss Benvenuto's and our answers, but that I could not discuss the questions to be discussed here. For as Clifford Truesdell never tired, and rightfully so, of stating, Newton's concept of force, which he called his greatest creation, is the first cardinal point of mechanics. The parallelogram law is the mathematical side of this concept, so that this question has a deep philosophical significance too, and I would love to have heard Benvenuto's opinions about this.

But then Prof. Corradi sent me another article by Benvenuto, one even more intriguing and fascinating, entitled "L'ingresso della storia nelle discipline strutturali" [Benvenuto 1988]. This paper comes very close indeed to accomplishing the task of finding in the history of the arts testimonies to the prehistory of a certain field of science.

In the background of Benvenuto's discussion stands the relationship between science, as represented by mechanics and engineering, and the arts, as represented by architecture. To be sure, he has here, above all, practical problems in mind, problems of restoration of old buildings. In his own words,

> La maggior spinta è venuta ... dall'esigenza di formulare plausibili interpretazioni del comportamento meccanico di strutture e materiali da costruzione che da millenni accompagnano la civiltà umana, come le murature, le volte, le cupole [Benvenuto 1988: 7].

So, like me, he was trying to find a bridge between the arts and science, but then he was looking at this bridge from the other end as well, as for example when he speaks "della scelta operata da Palladio per una integrazione della Storia dell' Architettura con la Storia delle Scienze che da sempre hanno sostenuto l'arte del fabbricare" [Benvenuto 1988: 7].

The cardinal point of the paper is a long quotation from the work of the German philosopher Arthur Schopenhauer [1819, III: 43]. He makes a series of comments in support of the philosopher's thesis:

> La lotta fra il peso e la rigidità costituisce... l'unico tema estetico dell''arte in architettura; far risaltare tale contrasto nel modo più vario e più evidente: questo è il suo ufficio [Benvenuto 1988: 9].

And further,

> ...frenando [le forze] col deviarle; così prolunga la lotta, e rende visibile sotto mille forme svariate lo sforzo infaticabile delle due forze nemiche. Abbandonata alla sua tendenza naturale, tutto l'edificio verrebbe a formare una massa compatta premente ... sul

suolo, su cui lo spinge inesorabile il peso... La rigidità invece...oppone a tale sforzo un'energica resistenza... Quindi la bellezza di un edificio consisterà nell'evidente adattazione finale di ogni parte [Benvenuto 1988: 9].

Benvenuto then analyses what he calls,

la traccia "estetica" dei tre principali temi della mecanica strutturale: il primo è... la "lotta" tra la "tendenza naturale" e l'energica resistenza; il secondo ...sono... le "vie tortuose" per quali l'architettura offre una manifestazione mediata della gravità; il terzo infine è ... la "finalità immanente" che direttamente "si referisce alla statica dell'linsieme" conferendo coerenza e teleonomica armonia ad ogni singola parte [Benvenuto 1988: 9].

From this basis Benvenuto sketches a program, especially for architectural archaeology, but as I am not competent in this field, I shall not follow him further in this direction. However, what strikes me, although Benvenuto does not ask it himself, is the following question: Which period and which style are evoked by Schopenhauer's conception and Benvenuto's analysis of them? Surely neither the Greek nor the style of the Renaissance, where symmetries, distinguished geometrical figures, proportions etc., when combined, lead to the harmonious, ideal building. Nor is it the Baroque, where such a *lotta*, or struggle, is usually carefully concealed, for instance with refined artifical painted architectures. And it was with these two styles that, naturally, Benvenuto was mostly concerned.

But Schopenhauer's words do indeed evoke the Gothic architecture of the Middle Ages, especially in France and Germany, and the Gothic is, indeed, the only style explicitly named by Schopenhauer in his text. I often wondered which geometric laws play a decisive role in Gothic architecture, the way they do in other styles. No doubt there *are* such mathematical laws in the Gothic style: just look at a cathedral and especially its plan, which is the result of an enormous imagination, punctiliously organized, even in the smallest details. And what is mathematics but organized imagination? Benvenuto's article suggests this to me, although he does not say it: it is not geometry, but her sister, or if you prefer her daughter, mechanics, that plays a decisive role here. I prefer not to say "statics" and would even prefer to say "dynamics", in the sense of the science of forces (*dynamis*) in equilibrium. And lest I should be misunderstood, I do not mean here "dynamic" in a vague philosophical sense. Exactly as "geometry", when we speak of the art of the Renaissance, means the geometry of the textbooks, i.e., Euclid's, in this instance I mean the mechanics of the textbooks: the science not only of the shapes of bodies, but of their interaction. And we know from Truesdell that, during this period, problems of resistance and elasticity were investigated. I shall quote him in a moment.

But in place of the circles, the squares, the proportions of figures, of what exactly must we think here? How do the dynamic ideas guide the builder and materialize into a beautiful building?

It is here that Truesdell comes in. He showed that Jordanus de Nemore had conjectured, if erroneously so as we know from Jacob Bernoulli, that the "elastica", i.e., the curve described by a bent beam is a circle [Truesdell 1960: 19]. Now, several people have told me that the Gothic pointed arch is composed of two circular arcs. But aesthetically these circles do not serve the same purpose as in classical architecture: they are not meant to be perfect *geometric* curves, but rather perfect *mechanical* curves. And thus Truesdell assures me, that what Benvenuto suggests to me goes in the right direction!

These are the curves that guarantee, as the pessimistic Schopenhauer could have put it, "for a building, whose height is dictated, the greatest possible security."   A mathematically-equivalent, yet optimistic formulation, which expresses, even better, the aim and boundless ambition of the builders is "given a dictated amount of money, material and thereby guaranteed security, these curves permit to build under these conditions the highest possible cathedral." Thus the strong vertically-thrusting drive that we observe and experience in a Gothic cathedral.

It would be lovely to continue this dialogue with Benvenuto on the *estetico*, but that would necessitate more space than is allotted to me, and frankly, as you guess, at the moment I am not yet fully prepared for this.

Before going over to Clifford Truesdell, let me add one more quotation from Benvenuto's article, about the role of historiography:

> Forse non è tanto un obiettivo storiografico ciò che sostiene questo interesse, quanto piuttosto la consapevolezza che un'approfondita conoscenza e un'attenta rimeditazione sul passato sono oggi condizione necessaria per un reale avanzamento della ricerca [Benvenuto 1988: 11].

Truesdell: mechanics and the art of historiography

When Clifford Truesdell entered the field of history of science, he was already the well-established author not only of many publications in various fields, but especially of many handbook articles. And furthermore, together with Walter Noll, he had formulated what may be called a new, deeply-structured mechanics of continua, which in their hands through the use of Hilbert's axiomatic techniques had become a totally unified domain.

This proved to be the best conceivable preparation for the future historian: equipped with this powerful armour he now turned more and more to the history of classical physics. Indeed, he could now see and display the historic development in perspective, step by step, the new ideas, the fruitful ones and the failures, the stimuli that they exerted, the accomplishments, always presenting clearly not only the context in which each scientist worked, but also "what he was up to." Combined with an intimate acquaintance with the original sources, he was able to assign to every question the exact role it had played, as well as to every author his proper place in that big stream. He showed how it all came together into what we know and understand today. Thereby often, with a few strokes, he could sketch the portrait of each actor in this epic and make visible his merits; of the great ones as well as the many forgotten and overlooked, unlucky ones. And above all, he could display the greatness of the field, whose history he was telling. I recall here especially his Introductions to Euler's work in hydrodynamics and in the theory of elasticity [Truesdell 1955; 1960].

The Introduction to Euler's hydrodynamics arrives at its summit, when Truesdell shows how, after Newton's work, Daniel Bernoulli through his equation unifies hydrostatics and hydraulics, John Bernoulli applies Newton's concept of force to fluids, d'Alembert reformulates this science by a field description and together with Euler introduces into this domain partial differential equations. Eventually then Euler, after having recognized the importance of what we call today an "inertial system", creates the new central notion: the "inner pressure". These discoveries allowed him to formulate the "Euler equations", the first field theory through which hydrodynamics and aerodynamics became unified! And thereby, as Truesdell showed, he opened the way to Cauchy's definitive formulation of the basis of the theory of elasticity.

The Introduction to Euler's works in the field of elasticity is a monumental treatise that, in its main part, traces its subject from Galileo to Coulomb. In the Prologue, which sketches its prehistory of more than 2000 years, he writes about the Middle Ages, and especially about the book written during the Gothic period entitled *Theory of Weight* by Jordanus de Nemore, whom I mentioned,

> ...*remarkable it is, Western in spirit, and ambitious beyond anything in the Greek and Arab tradition. The seventeen propositions on fluid flow, resistance, fracture and elasticity are all original....[Jordanus's] attempt at a precise argument to prove a concrete result in a domain never previously entered is of splendid daring* [Truesdell 1960: 18].

The main story is long, complicated and intertwined, and I cannot go into it here. It may suffice to say that, thanks largely to this Introduction, the history of the theory of elasticity, although the most complex part of the mechanics of continua, has become today perhaps the most carefully investigated one.

You will have noticed that often when I speak about Truesdell I use the word "concept". Indeed, he never tired of stressing the importance of the concepts used in science, especially of the concept of force, and it is from him that I learned, alas only very late in my career as a teacher, the importance of the role played by concepts. The concept is the mediator between the world of mathematics and the world of the senses, which in the end makes a science of the laws of nature possible.

This role is displayed and discussed in the book that Truesdell wrote with S. Bharata, *Classical Thermodynamics as a Theory of Heat Engines* [1977]. This is certainly his most important contribution to the art of teaching, from which every teacher can profit. It should be mentioned that Truesdell stresses here that for each theory presented to the students the teacher must indicate from the start the limits that the basic assumptions impose on it.

Conclusion and outlook

You may ask now, "What did the two men, Clifford Truesdell and Edoardo Benvenuto, have in common?"

First, both were great teachers, if we consider this word in its largest sense. I will mention here only some of the things that resulted from their parallel activities. Truesdell's work as a historian, especially of the 17th and the 18th centuries, is continued by two recent books written by Giulio Maltese on the history of mechanics from Newton to Euler, written in part in Genoa in close contact with Benvenuto [Maltese 1992, 2002]. These two books, whose author has become a foremost scholar in his field, are, I suspect, the first ones that profit fully from Truesdell's work and are a continuation of it.

Truesdell's health did not permit him to edit parts of Daniel Bernoulli's works, as he had hoped, and this task is now in the hands of Prof. Gleb Mikhailov. But the Bernoulli Edition, as it stands today, is unthinkable without his ideas, his experience, his unfailing advice, also as regards the beauty of the volumes and, most importantly, his constant encouragement. And thanks especially to Mrs. P. Radelet, the school of Genoa founded by Benvenuto is now, together with Prof. Maltese, collaborating with this enterprise, so that the efforts of both men do come together here.

And finally, both men were great lovers and connoisseurs of the arts, and indeed experts. No need to tell this of Edoardo Benvenuto here in Genoa. But I may add a few words about Truesdell. For in this field too his knowledge and erudition were stupendous. In Baltimore he had many friends among a group of craftsmen, who worked with him on the decorations of his stately home, where both he and his wife Charlotte, herself a musician, organized concerts played by artists on instruments built after the ones used in his beloved eighteenth century, the 'Age of Reason', as he lovingly called it.

Truesdell had a great sensibility not only for music, but also for the beauty of scientific texts, admiring the organisation of the matter presented, the transparency of a proof and the beauty of the mathematics or the mechanics itself, especially in the works of Euler, whom he compared to Bach and Mozart.

No wonder then, that as Charlotte Truesdell told me less than two weeks ago, that Benvenuto and Truesdell knew, liked and respected each other and that Truesdell (and knowing Mrs. Truesdell as I do, she as well) assisted Benvenuto with the English translation of his book.

Thus both men whom we honour today taught us that science, if presented by great men, can be beautiful too. When presenting its history this is much harder, but again both Benvenuto and Truesdell show us that this is possible: may their example excite and guide many followers!

Acknowledgments

It is a pleasure to thank Prof. M. Corradi as well as Prof. Becchi and Prof. Foce for their kind invitation to participate at the Symposium, and for their constant and generous hospitality. I am indebted to my wife for linguistic advice and to Mrs. Kim Williams-Pacini for having burdened herself with the edition of the Proceedings.

References

BENVENUTO, EDOARDO. 1985. The parallelogram of forces. *Meccanica*, vol. 20, no 2.

——. 1988. L'ingresso della storia nelle discipline strutturali. *Palladio* no 1.

MALTESE, GIULIO. 1992. *La storia di "F = ma". La seconda legge del moto nel XVIII secolo.* Florence: Olschki.

——. 2002. *Da "F = ma" alle leggi cardinali del moto: sviluppo della tradizione newtoniana nella meccanica del '700.* Milan: Hoepli.

SCHOPENHAUER, ARTHUR. 1819. *Die Welt als Wille und Vorstellung*, Leipzig.

SPEISER, ANDREAS. 1927. *Theorie der Gruppen von endlicher Ordnung*, 2. Auflage, Berlin: Julius Springer. pp. 87 - 96.

SPEISER, DAVID. 1976. La symétrie de l'ornement sur un bijou du trésor de Mycènes. *Annali dell'Istituto e Museo di Storia della Scienza di Firenze*, Anno I, Fascicolo 2.

——. 1987. Bernoulli, Mechanics and Restoration. Pp. 6-23 in *Die Werke von*

Daniel Bernoulli, vol. 3. Basel: Birkhäuser.

——. 2000. Architecture, Mathematics and Theology in Raphael's Paintings. Pp. 147-156 in *Nexus III: Architecture and Mathematics*, Kim Williams, ed. Pisa: Pacini Editore.

TRUESDELL, CLIFFORD A. 1954. Editor's Introduction: Rational Fluid Mechanics, 1687–1765. *Leonhardi Euleri Opera Omnia,* 2nd series, vol. XII: vii-cxxv. Zürich: Orell Füssli.

——. 1955. Editor's Introduction. *Leonhardi Euleri Opera Omnia,* 2nd series, vol. XIII Zürich: Orell Füssli. (Cf. esp. pp. x - cv.)

——. 1960. The Rational Mechanics of Flexible or Elastic Bodies, 1638-1788. *Introduction to Leonhardi Euleri Opera Omnia,* 2nd series, vol XI (2). Zürich: Orell Füssli. (Cf. esp. pp. xviii-xix.)

TRUESDELL, CLIFFORD A. and S. BHARATHA. 1977. *The concepts and Logik of Classical Thermodynamics as a Theory of Heat Engines.* New York: Springer-Verlag.

INDEX OF NAMES